Wetwares

Edited by

Sandra Buckley

Michael Hardt

Brian Massumi

THEORY OUT OF BOUNDS

Wetwares

Experiments in Postvital Living

Richard Doyle

Theory Out of Bounds   *Volume 24*

University of Minnesota Press

Minneapolis • London

Published by the University of Minnesota Press
111 Third Avenue South, Suite 290
Minneapolis, MN 55401-2520
http://www.upress.umn.edu

LIBRARY OF CONGRESS CATALOGING-IN-PUBLICATION DATA

Doyle, Richard, 1963–
Wetwares : experiments in postvital living / Richard Doyle.
p.   cm. — (Theory out of bounds ; v. 24)
Includes bibliographical references and index.
ISBN 0-8166-4008-4 (alk. paper) — ISBN 0-8166-4009-2 (pbk. : alk paper)
1. Artificial life.   2. Life.   I. Title.   II. Series.
BD418.8 .D69   2003
113'.8—dc21
2002155306

Printed in the United States of America on acid-free paper

The University of Minnesota
is an equal-opportunity educator and employer.

12 11 10 09 08 07 06 05 04 03   10 9 8 7 6 5 4 3 2 1

For Amy and Jackson

Contents

Acknowledgments

EVEN THE MOST SCHEMATIC TRANSCRIPT of the voices and impulses that have contributed to the production of this book would take longer to write than the book itself, so, dear readers, let the impossible compression begin.

Various allies have helped keep the Doyle going and periodically becoming, including, but not limited to, Celexa, Wellbutrin, Goya Espresso, Spring Creek Watershed, Rothrock State Forest, and McCoy Natatorium. Thanks go to the National Security State: its idiotic drug war has vastly improved the genetics and potency of Cannabis Indica and Sativa and its various hybrids. Good work, fellas!

The infosphere has been almost as productive as these more familiar allies; this book was begun during the first outbursts of the World Wide Web, whose exponential growth is the locationless prime mover of this paltry production and its testimony to the body of an emerging ecstatosphere. Forget the global brain and give thanks to evolution for getting on beyond cognition...

Speaking of forgetting: Sounds have worked on teaching me how to listen, and some of them have included: DJ Spooky, everybody who has ever turned a turntable into an instrument of chaotic joy, the recursive sounds of feedback—all praise Hendrix—the Infernal Chrome Gods and WKPS.

Every one of my students, both undergraduate and graduate, have hacked UC Berkeley and Penn State into ecologies of learning for all of us: thanks, and if you learned how to read you will hear yourselves scattered through every page of this book.

Moolah: This book's writing and research was aided and abetted by the Institute for Arts and Humanistic Studies and the Research and Graduate Studies Office at Penn State University. Thanks for the support.

Academic disciplines: Can't live with 'em, can't get paid without 'em. Thanks to the following people for saying yes to disciplinary innovation, often at great personal cost: Stan Shostak, Louis Kaplan, Susan Oyama, Zssa Baross, Philip Thurtle, Kathy Woodward, Michael Fortun, Paco Rodriguez, Suzanne Anker, Michelle Murphy, Stefan Helmreich, Hannah Landaecker, Chris Kelty, Norton Wise, Brian Rotman, Evelyn Fox Keller, Steve Shaviro, Lynn Margulis, and Elizabeth Wilson. The Science, Medicine, Technology, and Culture research group at Penn State (Robert Proctor, Londa Schiebinger, and Susan Squier) has helped turn our IPO into a remarkable research site. And thanks to all who have kept the ecstatic traditions of rhetoric from disappearing into a monoculture. Francesca Royster and Jeff Nealon have done the most to make working in the Penn State English Department joyful. Jeff Nealon has taught me too much to give him only one mention, so mega dittos. Mike Begnal has been the best of all possible neighbors and keeps the old school fires burning. Richard Morrison and Pieter Martin of the University of Minnesota Press get thanks for transforming my screed into something like a book, and Jennifer Smith helped me through the darkest hours of revision. Shouts to Theory Out of Bounds for making it possible to get between such lovely covers.

Thanks in advance to all readers for differentially actualizing *Wetwares*. Everybody else knows who they are. Thanks and praise go out to Gaia, which, like it or not, includes the NASDAQ.

Z　　　　　　E　　　　　　R　　　　　　O

Welcome to Wetwares™, N.0

Death needs time for what it kills to grow in . . .

> William S. Burroughs
> "Ah Pook the Destroyer"

It is no longer time that exists between two instants; it is the event that is a meanwhile (un entre-temps): the meanwhile is not part of the eternal, but neither is it part of time— it belongs to becoming. The meanwhile, the event, is always a dead time; it is there where nothing takes place, an infinite awaiting that is already infinitely past, awaiting and reserve. This dead time does not come after what happens; it coexists with the instant or time of the accident.

> Gilles Deleuze and Félix Guattari
> *What Is Philosophy?*

Addicted to the Future

THE BOOK was virtually complete. A disturbingly lively sheaf often appeared in his dreams—as an item of conversation, an agent in the author's life, a proliferating growth or graft that pleasured itself, alloaffectively, by gnawing persistently but delicately on the torn contours of dream space—and it became impossible to conceive of a daily life immune from its effects. *Wetwares* had probably become irreversible, kneading and needing the future. "Life may not advance, but it expands."[1]

The author, nonetheless, was not always so sure. There was always some story. Perhaps another chapter needed to be written, many sentences deleted. Much simultaneously must be given and abstracted, concepts abducted, deducted, and explained. The author had a sense that he had committed minor, absurd crimes, so he'd better get his story straight. Something repeatable, so that, under interrogation, it would spawn a monstrous labyrinth of questionable details.

It wasn't a matter of telling the story the same each time; far from it. Instead, the author sought to cultivate in his interrogators a massive response-ability, a response for this and a response for that. Even, especially, when he said nothing. In a work of fiction, he had encountered certain fabulously hot baths in Japan, where the least of movements would lead to a terrible and variable scalding. In such situations, hope resides in a continual response, allowing the future to arrive, dwelling multiply in the present through an almost molecular attention to the surroundings—Freeze! In such cases, knowledge is of course crucial, but so too is an aptitude for panic.

A Familiar Tale, or "Bodies That Splatter"

You wake up, for example, full of terrible and immediate cold. In a bath tub full of ice, in fact. You feel the wafting of a breeze in your gut, a raw open wound displacing an abdomen. Your mouth reeks slightly of tequila, a sensation of taste quickly overtaken by the agony of postsurgical trauma. You are in shock.

You scream for a while, of course. One needs a war cry, a slogan for such situations, a refrain prepared in advance for the job. Begin with onomatopoeia, as most howls become in their insistent tendency toward language. You might want to choose a scream for this purpose. You might even try "Yahoo!"

But a scream is, of course, a tremendous effort, so either calm down and cut to the chase or wait for exhaustion and vocal agony to set in. I recommend the former, but the latter is probably more fun to talk about later.

You look to your left and spot a postcard. Nobody bought it in a boutique. It's not even one of those free postcards functioning as a new form of distributed advertising, although it is an advertisement. It has visuals for easy reference—a crowd of twenty or thirty doctors in clean white lab coats and tiny red cursive names inscribed on their left pockets beam into the camera, all hoping for an improvement on their high school pictures.

Cut to an overhead shot. Zoom to a grainy close-up. "GREET-INGS FROM YOUR HEALTH CARE PROVIDERS" floats above the collected whorls of hair and skin in cursive Helvetica. You had better call them, the card improbably seems to suggest. An 800 number graces the back of the card.

You drag your shocked, swooning body to the phone. Where *is* the goddamn phone, not to mention your kidney? Fingers stab at buttons and the ringing begins until you recall your specially prepared refrain, a mantra offered here for the low purchase price featured on the book's cover. It's free if you are reading Xeroxes, stolen copies, or are just sitting in a megastore and executing the words right off the page.

"This is a hoax. This is a rumor. This is a rumor of a hoax."

A Familiar Tale, Part Two: Contact!

The book featured various bits of language that did much of the work in the story and allegedly were not confined to it. These were rhetorical softwares, unpredictable algorithms of textual hazard whose results were subject to change. They were part of the variable character of the book, its consistency of variation. Version N.0, he liked to call it, if only to register it with the proper authorities. Only the importance of nothing kept him from iterating the variable to N.N.

Each of the chapters—he could hardly avoid thinking of them as segments—featured a way of distributing bodies. Oh, there have been lots of ways to organize bodies—sick bodies, laboring bodies, criminal bodies, pleasured bodies, meditated bodies, medicated and buried bodies—but these were bodies whose morphology was uncertain and whose habitat was information. Sometimes, as with artificial life, there was a tendency to act as though there was nothing but information, but the book tried to map the ecology of these computer organisms and learned that their cultivation required more than software or hardware. He found these alife creatures to be quite seductive. Indeed, he sometimes thought that if biotechnology featured the production of organisms and enzymes most persuasive to the stock market, artificial life (alife) was an ecology in which the most seductive representations of life were cultivated with computers and robotics.[2]

Seductive to whom? Nietzsche raises his hand politely, all his sobriety helping to govern the arrival of delirious laughter: "Which *one*? Who *gets off* on machines?" Others had already done a remarkable job in detailing the communities of humans most transfixed by the production of creatures *in silico*. The demographic was white and yet psychedelic.[3] But even machinic seduction is itself a kind of possession, an overtaking that signals less the manipulative power of a self than its capacity for affective transformation. And seduction was hardly an *agency*—neither active nor passive, seduction involves a summoning of alterity, the cultivation of a familiar. "Relax," this algorithm of the familiar reads, "it's just a machine."

A Familiar Tale, Part Three:

Eighteenth-Century Hair, Pink Satin, and Nothing but Boots

What was a machine, again? Deleuze and Guattari insist that machines are fundamentally made of connection, a little bit of this and a little bit of that. A gun connects flesh and metal at a distance, networks of computers stitch together communities of chat and rumor. Even a refrain, a bundle of repetitions, is a machine. And yet these connections are also, paradoxically, cuts or slices:

> A machine may be defined as a *system of interruptions* or breaks *(coupures)*. . . . Every machine, in the first place, is related to a continual material flow . . . that it cuts into. It functions like a ham-slicing machine, removing portions from the associative flow: the anus and the flow of shit it cuts off, . . . the mouth that cuts off not only the flow of milk but also the flow of air and sound.[4]

The paradox of this formulation becomes less a logical problem than a machinic one as soon as we inquire into the conditions necessary to connection. Consider a Turing machine. The infinite tape (its "continual material flow") of such a machine is composed of binary marks such as "1" or "0." Computation emerges from the movement from state to state and the effects of such movements—one or zero, for example—on the computational head. Such a movement can be thought of as a connection or "string" of one state—one line of code—with another, but it is a linkage that proceeds from interruption—the end of one state and the emergence of another. The head of the Turing machine thus "cuts into" the flow of the infinite tape even as it connects binary marks to each other in its internal memory. These interruptions can undergo sudden changes in kind: this is less a problem of halting than of stammering. Thought of as the physical systems that they are, such machines are subject to periodic catastrophes or "criticality," a phenomenon I will no doubt return to later.

If machines are composed of cuts that connect—"Connecticut, Connect-I- cut!"[5]—it is sometimes useful to carve a distinction between two ecologies of machines: weapons and tools. A tool "prepares a matter from a distance, in order to bring it to a state of equilibrium or to appropriate it for a form of interiority."[6]

Tools are thus tendrils of repetition, bringing matter back to the self or its external doublet of order. No doubt these deployments wander an itinerary—each blow of the ax sculpts its edge otherwise, contingently[7]—but the ecosystem of the tool is organized "in" autopoiesis—the maintenance of a self by a self, what biologists Humberto Maturana and Francisco Varela characterize as the emergence of the inside and outside, a refrain that creates a territory.[8] Weapons, by contrast, incite not territory but deterritorialization: a horse, rather than being eaten, treated as energy for an interiority, becomes a vehicle, a way of linking one space to another, a shifting range or territory whose border is formed by speed. Perhaps the most compressed articulation of this machinic difference in kind reads: The tool summons repetition, weapons transformation. For certainly the tool also emerges out of an ecology—including the rhetorical practices through which a tool is bundled, the ways in which its capacities are rendered repeatable and available for feedback, such as documentation—but weapons, as this use of the plural suggests, come in packs—they differ even from themselves.

Stuart Kauffman's analytic typology of networks is also helpful in thinking about this differentiation of weapons and tools as respectively deterritorializing and reterritorializing. Kauffman has studied a class of "NK" networks, where N refers to the number of components in the system and K describes the connectivity of the network—the number of elements that are cross-coupled in the system. Kauffman seeks to model the levels of fitness that emerge from different ratios of N and K, and he argues that it is at the "edge of chaos" that highest fitness levels are clustered. Here the NK networks suggest that ecosystems with "too much" connectivity—whose limit is a "rugged" landscape, where each element is connected with each other and mutations are completely communicable—have low adaptive fitness. They also suggest that networks with low connectivity—where the number of connections is too meager—tend to be trapped in a static fitness location, unable to respond to even beneficial contagions in the network.

> Based on this reality, it is eminently plausible that Boolean networks in the ordered regime but near the boundary of chaos may harbor both the capacity to perform the most complex tasks and the capacity to evolve most adequately in a changing world.[9]

In this context, the weapons/tool distinction describes different styles of networks whose capacities for transformation differ in kind—they are qualitatively as well as quantitatively different. Indeed, in some sense they "harbor" more and less capacity for difference. Weapons, as agents of deterritorialization, introduce novel surfaces of contagion, opening up the system to new forms of connection as K goes through the roof. Tools, as components of territorialization, tend to insulate ecosystems from other habits and habitats, as K (connectivity) stays low enough to thwart most contagions at a distance, sheltering equilibrium.

Crucial to this differentiation is the character of the border between a weapon and other nodes in the network. As connectivity increases, the interface between a weapon and its assemblage becomes indiscernible—not eroded, imploded, or occluded, but in a state of such entanglement that any attempt to draw a distinction between weapon and network itself becomes a complex algorithm, an algorithm whose shortest description is probably itself.

William Burroughs's *Place of the Dead Roads* features Kim Carsons, shootist in training and pen name for Burroughs's frequent alter ego, William Seward Hall. For Carsons, the encounter with weapons entails less an acquisition of expertise than an itinerary of disaggregation. Carsons learns less to use the gun than to graft it—the condition of being a shootist is to become-gun. This is not a vague, violent imperative to "be" an object—it is difficult to avoid thinking here of Chevy Chase's *Caddyshack* command to "be" the golf ball—but instead involves a hospitality to an inhuman form, an integration of an alien entity into one's very habitat. Such an integration relies intensely on forgetting; one must be capable of responding to the new action of a body whose very eye is a node in a network of weaponry, a capacity linked to a forgetting or an undoing of the old arcs of eye, hand, and memory.

> Think of the muzzle as a steel eye feeling for your opponent's vitals with a searching movement. Move forward in time and see the bullet hitting the target as an *accomplished fact*. . . . I am learning to dissociate gun, arm and eye, letting them do it on their own, so draw aim and fire will become a *reflex*. I must learn to dissociate one hand from the other and turn myself into Siamese twins.[10]

To link to the assemblage of gun one must untie the knots between visuality, tactility, and temporality. Carsons experiences the muzzle as an orifice, a flowing, feeling deterritorialization of an eye that can now act at a tremendous distance, an action whose limit is contact with the future—*an accomplished fact*. This orifice composes a space of both activity and passivity, a locale of seduction rather than decision. BANG. The gun even periodically surprises him.

This connection to the weapon cannot exactly be forged on purpose. That would be a little bit like laughing on command—out of alignment, mis-

placed, mistimed. As with laughter, it was more a question of being capable of response, an undoing of one knot that makes another. Subjectivity, in stitches.

> I must learn to dissociate one hand from the other and turn myself into
> Siamese twins. I see myself sitting naked on a pink satin stool. On the left
> side my hairdo is 18th century, tied back in a bun at the nape of the neck...
> On the right side, wearing nothing but boots, I cover a nigger killing sheriff
> with my 44 Russian. This split gives me a tingly wet dream feeling like the
> packing dream, where I keep finding more things to go into my suitcases
> which are already overflowing and the boat is whistling in the harbor and
> another drawer all full of the things I need... [11]

To forge an alliance with a weapon, then, is not to treat it as a homuncular double of
the self. Nor does this acquisition of a technological familiar operate through the
simple addition of knowledge; a profound forgetting—whose hand is this, anyway?—
lets "them do it on their own." The intensity of this amnesis is thus not negative—
new knots and other "dos" are formed out of the undoings of the I. "Your hand and
your eyes know a lot more about shootin' than you do. Just learn to stand out of the
way.[12] Deleuze and Guattari summon this refrain in their discussion of the war machine. "Learning to undo things, and to undo oneself, is proper to the war machine:
the 'not-doing' of the warrior, the undoing of the subject."[13] This undoing proceeds
through the ecstatic labor of conjunction and interruption—Carsons is cleft not by
the impossibility of signification, but by the folds of an origami of subjectivity. A practice of folding—"on the left side... on the right side"—the capacity to split entails less
alienation than "dissociation," a disciplined differentiation whose transformations are
marked both by ellipses and conjunction, addition and hiatus: "*and* the boat is whistling
in the harbor *and* another drawer all full of the things I need...*and* another drawer
all full of the things I need...*and* another drawer all full of the things I need......"

A subject in flight—"and the boat is whistling in the harbor"—
Carson is a fugitive from identity itself as he breaks out into a multiplex of personae
divided spatially and temporally from themselves. Conjunction and ellipsis become,
in Burroughs's hands, machines for connection, an entanglement with another, even
if that other be silence... "Silence takes on the quality of a dimension here..."[14]
Entangled with the future, the ballistic collision of flesh and metal *becomes an accomplished fact* when the future itself is a *familiar.*

Familiars—a zone of interactivity between humans and animals,
"psychic companions" that blur the contours of human subjectivity—supplement Burroughs's analysis of weapons and their ecologies. Burroughs treats such a conjunction

as an inhuman hospitality.[15] Joe the Dead, Carsons's eventual assassin, has the strength to be affected by animals, the capacity to respond to a cutaneous blur of species:

> Cats see him as a friend. They rub against him purring. He can tame weasels, skunks and raccoons. He knows the lost art of turning animals into a familiar. The touch must be very brave and very gentle.[16]

Cultivating a familiar requires a certain touch. A friendship of great rigor, it offers not fusion but transduction, as familiar and friend acquire a common surface, each attempting to burrow into the other. "They rub against him purring." Through repetition—stroke, stroke, stroke, stroke—skin and fur become a multiplicity—neither human nor animal, but an ecstatic smearing of both. The inside and outside of each body acts less as a border as it becomes a zone of intensities. "More" and "less" characterize multiplicities whose accumulations and distributions can periodically cross catastrophic thresholds—the collapse of a sand dune, a spiking stock market, the irruption that traverses a synapse, a phase transition. The future?[17]

As a refrain—"draw aim and fire will become a *reflex*"—the arm/eye/gun complex shrugs off its relations to distance, perspective, and light and becomes a machinic *feeler*, trolling for interiority, "a steel eye feeling for your opponent's vitals." Treating the very organs of his opponent as a tactile surface that eludes its character as an "inside," Carsons enables a very specific contact with the future—death. Making *contact* with the future becomes less a metaphor than a rigorous, proleptic hospitality, a welcoming of a technological ensemble—gun, arm, eye. This techno-ecology is forged out of cuts, a severing of connections that erases old networks of reflex and solicits new architectures of flesh, metal, and time. This undoing is learned, at least in part, through a welcoming of even technological familiars that smears the contours of human subjectivity into a zone of connection. As with Kauffman's NK networks, increases in connectivity foster new clusters far from equilibrium, novel connections that emerge out of the new indiscernibility of inside and outside and the subsequent capacity for contact.

This solicitation operates as a tactile search (and destroy) engine for interiority and is announced in joy, that "tingly wet dream feeling" whose causal agency is the split itself: "*This split* gives me a tingly wet dream feeling like the packing dream" (emphasis mine). Transformed not by an entity but by that which resides *between* entities, a split, a hole, Carsons's training maps out the contours of a familiar alliance. Less the effect of an agency than a growth that proliferates between the refrain of agencies, Carsons's splitting emerges as an algo-rhythm of one identity practice and another, eighteenth-century hair, pink satin, and nothing but boots.

Panic Jam

By undoing the here and now, it opens the way to new spaces, other velocities. . . .

Questions, problems, and hypotheses bore holes in the here and now to end up in the

virtual world on the other side of the mirror, somewhere between time and eternity.

Pierre Lévy, *Becoming Virtual:*
Reality in the Digital Age

Alife creatures, too, were particular sorts of familiars, silicon grafts, a machinic becoming-animal. Lycanthropy for networks! They were sometimes allopoietic, ports or links to something other than either the maintenance of a self or its dissolution. They were links to panic, the order word for his discussion of alife. Around alife, people begin to talk and get carried away. More than simple computer models or simulacra of that old concept, life, alife seems to be an interface that produces the incessant questions: *How is something alive? When will I know?* Artificial life disturbs, continually rendering the border between life and nonlife, flesh and machine, seductively uncertain. The very border of the flickering alife creature—the morphology of its phenotype—is in constant question, in stitches. On one side, the creation of organisms iterable enough to move from computer to computer, capable of being copied across networks, undoes the monopoly of carbon on living systems and extends the franchise of vitality to an already uncannily mobile machinic phylum. On the other tentacle, it deterritorializes life itself, as life becomes an explicit virtuality, placeless and yet distributed, ubiquitous. It becomes possible that everything is alive— panic. Nowhere in particular, indiscernible, life outsources the labor of representation to the newly excited surfaces of computer screens, techno-familiars who are poster creatures for vitality.

   "Panic is creation."[18] If panic qualifies as at least one actant in a creative ecology, its efficacy resides in a facility for rendering borders and screens indiscernible. In a panic, the surfaces of the social become slippery—a capacity for movement arrives as a variable fluidity is discovered in the very sturdiness of things. A plunging stock, a stampeding crowd, a spitting, cascading flood of lava all index the sudden arrival of locationless, distributed movement. Panic does not entail a fundamental loss of control, but instead occasions the emergence of a new, incommensurable order. There is a logic, or at least a rhetoric of the stampede, the pack, the swarm—a qualitative difference emerges as a docile crowd becomes a mob incited by rumor. An orgy breaks out in multiplicity, "to lose themselves in the alterity of the collective."[19]

Paradoxically, such changes in kind are often heralded by a sudden blockage—a computer crashes, a pond freezes, a creek pools up against a downed limb, an elevator is stuck between floors. Such blockages form new contours for repetition—lap, lap, lap—and, in their very restraint, forge novel zones of excitation, whirlpools, eddies, turbulence of air and prices. Sociologist Michel Maffesoli describes the flows of a generalized *orgiasm* in terms of a refraining, the iterative constraint of a "bridle": "The same thing holds for spending as holds for violence: bridling it in its expression is in fact encouraging its perverse and exacerbated irruption."[20]

And yet, to paraphrase Nietzsche, the limb does not *refute* the creek. Rather, this repetitive "bridling" is the very substrate of emergence or "irruption." Researchers such as Cairns Smith and Stuart Kauffman write of a constitutive clumping that occasions the arrival of order at the edge of chaos and enables phase transitions of all kinds.[21] "Refrain"—etymologically, both repetitious ditty and bridle—itself clumps or clusters together divergent, qualitatively distinct rhetorical characteristics—the capacity to be repeated and the capacity for interruption. Writers such as Judith Butler have noted the productive effects of iteration in the constitution of gender and identity but have been perhaps less attentive to the creative emergence of inhuman, nonlinear transformations through such repetition—rumors, sand dunes, traffic jams, new capacities for subjection installed less by lack than by novelty. As the substrate of both habit and habitat, though, refrains compose a crucial and productive component in the "exacerbated irruption," not a reaction against it. Instead, such repetitious blocks form novel and robust flows in ecologies of catastrophe, what Maffesoli links to Goethe's notion of "mobile order . . . it is always by blocks, by ensembles, that things and people are moved."[22] In short, the bumper sticker for such an inhuman politics reads: Provoke swarms, forget coalitions.

What sorts of events and organisms do such novel flows select for? Put another way, what sorts of ecologies cultivate such contingencies and communities? As an engagement with multiplicity—one is "beside oneself"—encounters with panic, rather than reactions against it, are fostered by a counterintuitive ethos of *subtraction*. Deleuze and Guattari write of the need to subtract the One from the unknown or *n*.

> The multiple must be made, not by always adding a higher dimension, but rather in the simplest of ways, by dint of sobriety, with the number of dimensions one already has available—always $n - 1$ (the only way the one belongs to the multiple: always subtracted).[23]

This subtraction operates by amplifying and extracting the capacities of the anomalous to be deterritorialized, the line of flight. The cutting of a diamond, for example, must follow the singular character (shape, lines of cleavage, flaws, color) of the stone and determine its heterogeneity to effect the most ecstatic dispersal of light, a dispersal whose intense dislocation—where is a flicker?—nonetheless localizes value most effectively. It is by dint of this "value" that a vector distributing diamonds emerges, a deterritorialization whose attraction operates on the arms and lungs of the miner and the glittering skin of a starlet. This attraction is itself cultivated through a subtraction or a "flattening": "the diamond must have flat surfaces called facets to act as prisms before it can release its brilliance and dispersion."[24] A cartel to block sales of an incredibly common stone doesn't hurt matters, either.

This cultivation of the anomalous, "itineration," emerges from repetition, a doubling. Facets cut and echo each other, not as "reflective" surfaces—what do they echo but each other, light itself?—but as the iteration of light on light. Composed of repetition—grind, grind, grind—the very cutting and polishing of a diamond demands a strict "following" of the particular qualities of each stone. As with gem cutting, Deleuze and Guattari argue that primitive metallurgy is necessarily an "ambulant science," essentially itinerant in its practice.

> One is obliged to follow when one is in search of the singularities of a matter, or rather of a material, and not out to discover a form . . . when one ceases to contemplate the course of a laminar flow in a determinate direction, to be carried away by a vortical flow.[25]

Cleaning a stone or a dish, even reading about it—scrub, scrub, scrub—demands such an itineration: You missed a spot! This itinerant repetition dislocates or distributes identity through its proximity to an indiscernible doubling. Panic: *Where was I?*

And "to be carried away" by such a flow, one must often instigate a blockage.[26] If the multiple must be "made," it is because its unfolding demands a *refrain*. Through continual repetition, for example, Carsons "dissociates" and learns to "stand out of the way" of gun, hands, and eyes. The univocal identity called "self" gets blocked, refrained, flattened, and this blockage emerges precisely through an attention to something else, the acquisition of a familiar. Deleuze and Guattari suggest that itineration or "an exploration by legwork" jams point of view and blocks the self's function as a reference or reproduction:

Reproducing implies the permanence of a fixed point of *view* that is external to what is reproduced: watching the flow from the bank. But following is something different in kind from the ideal of reproduction.[27]

The discipline of itineration entails an ecstatic but disciplined receptivity to the turbulent outside where "the self is diluted into a more viscous and confusional entity."[28] *They rub against him purring*. This dilution, though, is a question of intensity—a difference in kind—rather than loss. Following involves a drifting, entangled complicity "in an objective zone of fluctuation that is co-extensive with reality itself."[29] Neither suffering nor enjoying a static "point" of view, the itinerant continually acquires novel capacities to flow, the strange capacity for invasion that biologist Lynn Margulis links to the emergence of endosymbiosis and Deleuze and Guattari refer to as thought. These capacities for difference often emerge precisely out of a blockage of an autopoietic self—a self made out of the difference between inside and outside—and the consequent emergence of a tangle or a "grapple" ("A thought grappling with exterior forces instead of being gathered up in an interior form, operating by relays instead of forming an image").[30] This grapple, as in the case of Carson, need not simply involve "biological" entities—familiars could be technological, even textual. "Kim could feel the phantom touch of the lens on his body."[31] What had been distant—flesh, lens, gaze—acquires a common surface, not a static bonding but an undulating, variable flow, "light as a breath of wind."[32] The familiar involves, therefore, an action at a distance, but it is an action whose effects transpire in no given location—neither eye, lens, nor body—but as a virtual, "phantom touch" or a throbbing "medium":

> Held in a film medium, like soft glass, they are both motionless except for the throbbing of tumescent flesh... "Hold it!"... CLICK... For six seconds the sun seems to stand still in the sky.[33]

An act of distribution, the acquisition of a familiar proceeds only when identity resides in no specific place or time whatsoever. Computer scientist Pierre Lévy characterizes this "neither here nor there" quality as a fundamental attribute of the virtual, an ecology where, like many practitioners of the new sports of the between, entities are nowhere in particular: "the surfer or parachutist is never entirely *there*."[34] Less a point of view than a rhythmic, ecstatic capacity for difference, humans enter into relations with familiars and access newly contacted surfaces: holes sprout in what had been experienced as wholes.

Just You Watch!

Psychologist James Hillman, in his singular treatment of panic, notes that even the itinerative act of observation can open up such holes, as the rigorous pursuit of attention creates vertiginous whirlpools of affect that suck the observer into the suspension of infinite regress:

> Pan appears again and again as an *observer*. There he stands, or sits or leans or crouches, amidst events in which he does not participate but where he is instead a subjective factor of vital intention. Wernicke says he serves to awaken the interest of the onlooker, as if when we look at a painting with Pan in its background, we are the observing Pan.... Within the physical intensity of Pan there is a physical attentiveness, a goat's consciousness.[35]

Pan, appearing as an observer, "awakens" the viewer with a start. Suddenly, we are more than lookers—our observations are doubled, mimed by a human/animal hybrid who does nothing but make disturbingly visible the mechanisms of the very *outside of paint*. Pan, in his visibility, renders indiscernible the interior and the exterior of the painting. The viewer is awakened into the operation of flesh and all its sensitivities—"attentiveness"—which instantiates the work of art. The fundamental complicity of observer and observed is experienced as a vertiginous dissolution of the ready-made differentiation between the interior and exterior of the painting, the interiority and exteriority of a body. Gilles Deleuze finds this production of indiscernibility to be a particular feature of Francis Bacon's extraordinary canvases of meat:

> What Bacon's painting constitutes is a zone of indiscernibility, of undecideability between man and animal.... Bacon pushes this to the point where even his most isolated figure is already a coupled figure, man is coupled with his animal in a latent bullfight. This objective zone of indiscernibility is the entire body, but the body insofar as it is flesh or meat.[36]

Coupling with Pan in the very act of observation, observers find themselves awash not in reflexivity—I am watching a watcher watch me while watching—but an "infinite" refrain whose necessary finitude immerses one in an abyss of indiscernibility, the catastrophe of flesh—I am watching a watcher watch me watching it watch, I have lost count, where was I? *We have become the observing Pan.*[37] Less a failure of observation than the production of and capacity for indiscernibility—becoming-Pan— the appearance of Pan and the apprehension of meat both provoke strange Möbius topologies of observation and identity, a panicked experience of ecstasis in which one is both inside, outside, and neither. Deleuze writes of the disturbing proximity of meat's tremendous difference, its testimony to suffering: "The immense pity that the meat entails."[38] This encounter with interiority's fundamentally folded character—what seemed to be one, autonomous space now becomes a fractalled, crinkly "zone" of variation where inside and outside, proximate and distant, open a multiplex—cultivates affects of often pitiful contagion and complicity with an allegedly inhuman environment. Deleuze argues that, in the case of Bacon, this "coupling" provokes an ethos where the very borders between humans and animals—that zone of the familiar—become indiscernible: "Meat is the common zone of man and the beast, their zone of indiscernibility."[39]

Hence the "goat consciousness" alluded to by Hillman above—that surefooted, itinerative following of the tangled pathways of becoming, paths that are in some sense available only in their traversal or their often panicky "running." Such an itineration demands a "hole" in the usual operations of selfhood—the outside must be attended to with immense precision—this rock, that root, that paint must be responded to with foot, hand, eye, thought rather than consciousness. One less creates a line or a path than, in the hallucinogenic words of Henri Michaux, "becomes a line" in a sudden event of exteriority's indiscernible shrinkage or flattening of the self into a line.

> Becoming a line was catastrophic, but it was, still more unexpectedly (if that's possible), prodigious. All of my self had to pass through that line. And through its horrible joltings. . . . Metaphysics overtaken by mechanics. Forced through the same path, my self, my thought, and the vibration. Self a thought only, not thought becoming my self or developing in my self, but myself shrunk into it.[40]

Catastrophe and prodigy both resound with a sudden incursion by the future. Prodigy is the very sign of futurity, a mark that is the trace of the to-come, "if that's possible," while catastrophe recalls the fundamentally rhetorical fabric of the surprise of sudden arrival, what the *OED* renders as "the change or revolution that produces the conclusion or final event of a dramatic piece." More than space is smeared in this zone of indiscernibility between sign and future—the clean border between present and future becomes slashed, leaking into the sudden jolting of qualitative difference in that zone of variation present/future, becoming.

Michaux's prodigious and yet miserably vertiginous becoming—a flatness of line that yields neither inside nor outside, neither present nor future—

extends to the precise and thoroughly *malleable* actions of sensation, a work of remarkable physical complexity detailed by Paul Virilio in his discussion of Rodin. "In order to *sense* an object with maximum clarity, one must accomplish an enormous number of tiny, rapid movements from one part of the object to another."[41] This accomplishment, though, involves a distributed agency—neither here nor there but an itinerant, diffuse sensitivity that enables objects to traverse the senses. The event of observation is less a passive reception than an incessant exposure to a swarm, a hospitality to the multiple "parts" of an object. This hospitality demands both continual openings and closings, the blockages, flattenings, and exposed surfaces of a refrain. Michaux, this time in alliance with mescaline:

> As if there were an opening, *an opening like a gathering together*, like a world, where something can happen, many things can happen, where there's a whole lot, there's a swarm of possibilities.[42] (emphasis mine)

Neither passive nor active, this facility for observation proceeds through a "hosting" of sensation. Overtaken, the self is riven by fractures of connectivity, alliances with alterity that gather, cluster, and flock even as identity's surface is punctured with connections to the outside, an opening that is not an absence. Such alliances can often block "blockage"—the rigid distinction between inside and outside, present and future—and enable a swarm, an organized and distributed multiplicity.[43]

Hence itinerant observation—a blockage of the self and an opening onto a swarm of possibilities, "following"—provokes capacities for futures, futures that defy prediction in their "fundamental alienness."[44] Stuart Kauffman encounters this alien character of futurity in his discussion of algorithmic complexity, a principle of computer science that renders humans into slack-jawed observers of the future, contingency addicts for whom any transformation of the present can be encountered but not compressed.

> For vast classes of algorithms, no compact, lawlike description of their behavior can be obtained....If the origin and evolution of life is like an incompressible computer algorithm, then, in principle, we can have no compact theory that predicts the details of the unfolding. We must instead simply stand back and watch the pageant...If we demand to know its details, we must watch in awed wonder and count and recount the myriad rivulets of branching life and the multitudes of its molecular and morphological details.[45]

Kauffman refers here to phenomena that physicist Stephen Wolfram, among others, dubs "irreducible," algorithms that are likely to be the shortest possible description of themselves. Such algorithms or recipes require a very specific ingredient: futures. Outside of the actual instantiation of its code, the running of the program, observers can have no detailed foreknowledge of the system. "We must simply stand back and watch the pageant." Hence the actual computation—even counting—of complex algorithms demands, at the very least, software, hardware, and futurity. Such algorithms qualify as what Deleuze and Guattari characterized above as "an exploration by legwork" or itineration—there is no cutting to the chase. "They do not meet the visual condition of being observable from a point in space external to them."[46]

Such an entangled observation is thus more than passive; it involves a profound stoppage, a capacity to endure the meantimes through which the details of the future emerge. In Kauffman's example, we must even endure "awed wonder," the sublime incapacitation of self by the agonizingly gorgeous unfolding of complexity. But such endurance is more general than this instance of the sublime—Kauffman too, *as just such a complex algorithm*, can only, actively, wait. This active waiting is an exposure or a hospitality to the future: difference is hosted by a self that lives not simply off identity but through the sensitivity to and capacity for variation, allopoiesis. In Kauffman's example above, we must at the very least be capable of registering countable differences—we must be available for the sudden flow of surprise, "myriad rivulets of branching life and the multitudes of its molecular and morphological details." More than the arrival of information, surprise shocks with asignification, tearing configurations of meaning and fostering the emergence of new forms of subjectivity, "these ruptures of meaning that are auto-foundational of existence."[47] What happened?

Seduced by Science/Fiction

These ruptures, too, are anything but locatable—neither here nor there, they are less absences or nonbeings than bodies in the midst of becoming, what hole theorists Roberto Casati and Achille C. Varzi characterize as *immaterial bodies*. Of holes, they write:

> They are not parts of the material objects they are hosted in (though it is sometimes by removing a part of the host that a hole is created); rather, they are immaterial bodies, located at the surfaces of their hosts.[48]

These immaterial bodies are virtual companions or familiars—neither beings nor nonbeings, but promising strangers that emerge only in response to a hosting. Deleuze and Guattari write of the necessity of tearing such holes through clichés and opinion to enable the arrival of difference in the form of chaosmos: "A chaosmos, a composed chaos—neither foreseen nor preconceived...Art struggles with chaos but it does so to render it sensory."[49] Composed of sensational events, art hosts such openings in the form of affects such as indiscernibility.

> When Fontana slashes the colored canvas with a razor, he does not tear the color in doing this. On the contrary, he makes us see the area of plain uniform color, of pure color, through the slit. Art indeed struggles with chaos, but it does so in order to bring forth a vision that illuminates it for an instant, a Sensation.[50]

Fontana's cut of the canvas connects paint to a vision that transpires in a zone of indiscernibility. Is such a hole inside or outside the painting?

Technoscience, too, engages in such a struggle or a grapple—it is seduced by all that exceeds it: "science cannot avoid experiencing a profound attraction for the chaos with which it battles."[51] Sometimes this battle—rather than a *war*—involves a similar slashing tactic—a cut that connects, a rendering indiscernible between science and fiction, science/fiction.[52] Practices of chaosmosis—a welcoming neither foreseen nor preconceived—enable both ruptures and connections through which transformative networks of technoscience and wetwares sometimes emerge.

Maybe the book had become this sort of familiar to him, a graft that interrupted the host's sense of the interior of the world and its exterior, an orifice for action at a distance, distribution.

O N E

Representing Life for a Living

Panic . . . the sudden, intolerable knowing that everything is alive.

William S. Burroughs, *Ghost of Chance*

From Agents to Events: Distributing Life

EVERYONE KNOWS that in 1953 James Watson and Francis Crick diagrammed the structural and functional characteristics of the double helical molecule deoxyribonucleic acid. Much of the rhetoric of this remarkable achievement suggested that Life's secret had finally been uncovered, and that Life was therefore localized in the agency of genes. Watson, writing in his autobiography named for a molecule, put it this way: "In order to know what life is, we must know how genes act."[1] The action and manipulation of nucleic acids became the hallmark of a molecular biology that no longer analyses organisms but—in symbiosis with stock markets and venture capitalists—transforms them, generating life-forms without precedent.

And yet this localization of life onto genetic actors—"what life is"—has also enabled an astonishing distribution of vitality, one that allows us to speak of "artificial life," simulacra that are not simply models of life but are in fact instances of it. In short, life is no longer confined to the operation of DNA but is instead linked to the informatic events associated with nucleic acids: operations of coding, replication, and mutation. I will argue in this segment that the emergence of artificial life signals more than the liberation of living systems from carbon—it maps a transformation of the scientific concept of life itself, a shift from an understanding of organisms as *localized agents* to an articulation of living systems as *distributed events.* Not located in any particular space and time, living systems are in this view understood as unfolding processes whose most compressed descriptions are to be found in the events themselves—growing explanations.[2] This chapter will focus primarily on the challenges that such a transformation poses for our means of representing life in both scientific and extrascientific texts. I analyze both rhetorical milieus, not because such realms of discourse are equivalent, but because they can sometimes fruitfully borrow from each other. Indeed, as I hope to make clear, the rhetorical practices involved in such representations of life are potent allies in the transformation of life, as rhetorical "softwares" become elements in the very network through which artificial life becomes lively.

Alife Makes Me Nervous

Alife, I must admit, makes me nervous. This wracking of my nerves is not the deep anxiety of any fear and loathing of machines, those alleged alienators of souls, labor, depth, bodies—in short "'technology." No, alife is like a joke that I just can't seem to get. You know the feeling. Oh. I'll admit that some of the creatures are, well, cute, that they scamper across my screen, seemingly out of control, on the same drugged electricity as the Energizer battery bunny. Sometimes I'll get all enthused about a simulation, sitting at my desk while I simulate work, and fetch a colleague from across the hall. "Look!" I'll say." "Look! IT'S ALIVE!" and point at the swirling, flickering, flocking pixels. "It's alive!"

There just seems to be no convincing him. I've even tried the old rhetorical ploy of implicating my neighbor in a simulation. With LifeMaker, a cellular automata program available on the web, I spelled out his name in "cells" and put the simulation on ultrafast. The cells swirled, flickered, and dissolved the writing and produced something more akin to a proliferating growth than a name. My colleague, a Joyce scholar, just looks at the screen, then at me, and deadpans "It's just language, rhetoric boy."

In these situations I realize that I am being called upon to justify my expertise, so I lean back in my chair, do my best impression of a professor, and provide some historical and cultural context.

"Just *information,* you mean." I say this as if I have trumped him, as if the distinction is itself so stuffed with *information* that the scales should fall from his eyes. "But since Erwin Schrödinger's articulation of the genetic substance as a 'code-script' in 1943, life itself has gradually been conflated with *information*. The trajectory is long and complex—from George Gamow's 1954 discussion of the "diamond code" scheme for the translation of DNA into proteins, Jacques Monod and François Jacob's research on induction and the genetic "program," to the recent human genome initiatives and their mapping and decoding of the "Book of Life"—but suffice to say that from the perspective of many contemporary biologists, life is just an interesting configuration of *information*. Biologist Richard Dawkins, who you may have read about in the *New Yorker*, has claimed that we are nothing but 'lumbering robots,' vehicles for the propagation of DNA. We're like the host, it's the parasite. Or maybe it's the other way around."

"I still don't get it. Which part is alive?"

At this point I despair, because, I have to admit, it starts to feel like I am explaining a joke, a task that is neither fun nor funny. I can't really pull off the knowing, laughing: "Oh, you don't get it. Well, if you don't *get it*, I can't really explain it. It would just take too long." At least, I can't *say* any of that. So all I can really do is imply it, suggest through my silence that it's a generational, theoretical thing. Perhaps he doesn't have the secret poststructuralist alife decoder ring, I suggest through my silence. "I'll give you a copy of this segment I am writing about alife when I am done," I say. "Maybe that'll help."

So alife makes me nervous because it seems to be inarticulable in some way. And periodically, I don't get the joke. But it also makes me nervous because I can't simply write it off. Because there is something uncanny about alife. It's a creepy doubling of something that no longer appears: "Life."

"Life," as a scientific object, has been *stealthed*, rendered indiscernible by our installed systems of representation. No longer the attribute of a sovereign in battle with its evolutionary problem set, the organism its sign of ongoing but always temporary victory, life now resounds not so much within sturdy boundaries as between them. The very success of the informatic paradigm, in fields as diverse as molecular biology and ecology, has paradoxically dislocated the very object of biological research. "Biologists no longer study life today,"[3] writes Nobel Prize winning molecular biologist François Jacob, "they study living systems." This "postvital" biology is, by and large, interested less in the characteristics and functions of living organisms than in sequences of molecules and their effects. These sequences are themselves articulable though databases and networks; they therefore garner their effects through relentless repetitions and refrains, connections and blockages rather than through the autonomous interiority of an organism. This transformation of the twentieth century life sciences, while hardly homogeneous and not univocal, marks a change in kind for biology, whose very object has shifted, become distributed.

Consider, for example, the Boolean networks of Stuart Kauffman's research discussed above. Kauffman argues in his work that such nets—networks of buttons threaded to each other in a random pattern—display autocatalytic configurations after a phase transition when the ratio of threads to buttons reaches .5, after which "all of sudden most of the clusters have become cross-connected into one giant structure."[4] Kauffman and others find this deep principle of autocatalytic nets to be suggestive of another transition—that at the origin of life:

> The rather sudden change in the size of the largest connected cluster of buttons, as the ratio of threads to buttons passes .5, is a toy version of the phase transition that I believe led to the origin of life.[5]

The paradox of such a formulation emerges around the dependence of this remarkable model on the "sudden" quality of the transition. Precisely the power of such a network—a persuasiveness that would link surprise to the very emergence of life—renders a difficult problem for a rhetoric that could narrate such a "sudden change"—It's alive! How to articulate such a parallel event, an event in which difference emerges not in a serial, one-after-another story, but all at once, all of a sudden? To visualize this change in kind, Kauffman's own visual rhetoric has recourse to a flicker or a blinking when he constructs a Boolean net composed of bulbs that light up based on their logical states:

> I will assign to each light bulb one of the possible Boolean functions... AND function to bulb 1 and the OR function to bulbs 2 and 3. At each tick of the clock each bulb examines the activities of its two inputs and adopts the state 1 or 0 specified by its Boolean function. The result is a kaleidoscopic blinking as pattern after pattern unfolds.[6]

Crucial to representing such a networked understanding of life is a semiotic state that, like Levy's parachutist, is neither here nor there—the flicker of such a "kaleidoscopic blinking" signals less the location of any orderly pattern than its status as a fluctuation—a configuration that emerges precisely between locations rather than "in" any given node of the network. The blinking—neither on nor off, but the difference between on and off—signals the capacity of each specific type of network— such as Kauffman's K = 2 networks, where each bulb is connected to two other nodes—for orderly patterns. In short, even in models of that allegedly singular event, the origin of life, articulations of life involve a multiplicity—multiple nodes—whose "liveliness" emerges between locations.[7] This notion of life relies less on classical conceptions of autonomy than on a rigorous capacity for connection, orderly ensembles representable only as transformations, flickering patterns unfolding in space and time.

This makes alife's claims concerning the vitality of virtual organisms all the more perplexing. For if life seems to have disappeared as a sovereign entity and joined the ranks of all those other relational attributes—economic value, for example—then it seems odd that it should reappear, so visibly, on my screen.[8] It's enough to make one believe in time travel. It's as if computers, with the right softwares, could travel back to a past when life was an autonomous attribute of organisms, capture it, and display it on the screen.

Heterodox theoretical biologist Marcello Barbieri has argued in this context that organisms have always already been networks. Moving beyond a rendering of the organism into the duality of "genotype" and "phenotype," Barbieri argues that living systems must be understood as a tripartite ensemble of *genotype*, hereditary information primarily although not exclusively born by DNA; *ribotype*, the swarm of translational apparatuses that transform DNA into the tertiary structures of folded proteins; and *phenotype*, the dynamic embodiment of these informations and their transformations.[9] Crucial to Barbieri's argument is the recognition that DNA "information" is necessary but not sufficient for the emergence of life; yet another translational actant is needed to transform the immortal syntax of nucleic acids into the somatic semantics of living systems.

By analogy, I want to suggest that alife, too, emerges only through the complex of translational mechanisms that render it articulable as "lively." The ribotype that transforms the coded iterations and differences of alife softwares into the lifelike behavior of artificial life is composed of, among other things, "rhetorical softwares." These rhetorical formulations—as simple as a newly coined metaphor or as complex as an entire discourse—don't "construct" scientific objects so much as they discipline them, render them available for scientific observation, analysis, and argument, much as the flicker of bulbs images Kauffman's autocatalytic order above. The rhetorical challenge posed by life that emerges out of networks goes beyond the ontological uncertainty that haunts artificial life—are they really alive?—and becomes a problem of articulation: How can something that dwells not in a place but in virtuality, a network, be rendered? Hence rhetorical problems haunt not simply the status of alife creatures, but their location.[10]

It is perhaps due to this uncanny distribution of life that the parallel rhetorical formulations of "localization" and "ubiquity" have particular force on artificial life, even as these effects are in tension. Rhetorics of "localization" suggest that some particular organism "in" or "on" the computer is "alive," thereby occluding the complex ecology of brains, flesh, code, and electric grids that alife thrives on and enabling the usual habits of narrative—an actor moving serially through a world—to flourish, as a more recognizable and perhaps seductive understanding of an organism as "agent" survives. At the same time, rhetorics of "ubiquity" provoke the possibility that, as in Burroughs's observation above, *anything could be alive*, leaving the observer continually alert to the signs of vitality in the midst of machines. It is this latter effect that produces much of the excitement of artificial life, a sublime incapacity to render the sudden, ubiquitous complexity of a phase transition into the sequential operations of narrative.[11]

All scientific practices are differently comported by their rhetorical softwares: my focus has been on the researches and insights enabled by articulations of organisms as extensions of "code." But alife is in a slightly different position with respect to its rhetorical components, as the actual *difference* of artificial life, as "life," is continually at stake. This crisis of vitality that pervades alife is not simply due to alife's status as a "simulation"; as I suggested above, alife merges out of a context in which quite literally, *life disappears*, as the "life effect" becomes representable through the flicker of networks rather than articulable and definable locales. As researcher Pierre Lévy writes, "Virtualization comes as a shock to the traditional narrative."[12] My challenge here will be to determine the specific rhetorical mechanisms that enable the narration and instantiation of some versions of artificial life, a wetware ribotype that makes alife such lively creatures at this moment.

As virtual organisms, alife creatures are not fake. Like all simulacra, they are copies without original, producing an effect not of reference—what would they *refer* to?—but of provocation, the uncanny feeling of familiarity in the unfamiliar realm of the computer screen. They double and *fold* the organic into the virtual, a hybridization of machine and organism that, inevitably, makes one laugh. We laugh nervously because while we are not in a state of Burroughsian Panic, all of the technological in Frankensteinian rebirth, full of life—no conspiracy involving a toaster, chainsaw, and a couple of CD players is in the cards—we nonetheless get the sense that indeed anything *could be alive*. This is the first element I'll assay in the ribotype of artificial life, a rhetorical ensemble that smears the borders between the computer and its environment, what we could call a silicon abduction.

Abducted by Silicon

If artificial life creatures, as actualizations of information, enjoy the burdens and benefits of vitality, they do so through the operation of what Charles Sanders Peirce characterized as "abduction." Peirce, a nineteenth century polymath contributor to mathematics, semiotics, and philosophy, formulated his theory of abduction in order to supplement the more traditional logical categories of induction and deduction. Scientific thinking, Peirce held, didn't always proceed via the clean operation of these categories. Kepler's discovery of the laws of planetary motion was among Peirce's favorite examples of a scientific practice that differed from these logical frameworks. Abduction, as a category of reasoning, is characterized by its reliance on an *absence:*

> An *abduction* is a method of forming a general prediction without any positive assurance that it will succeed either in the special case or usually, its justification being that it is the only possible hope of regulating our future conduct rationally, and that induction from past experience gives us strong encouragement to hope that it will be successful in the future.[13]

A missing term—one that may possibly arrive in the future—completes abduction's argument. The "possibility" that inheres in any specific abductive enterprise is tethered to the pathos of "hope," an encounter with the future without grounds but with calculation, anticipation, and a bit of desperation—"the only possible hope of regulating our future conduct rationally." The past, too, offers itself up as a support to abductive reasoning, but only in the form of the "inductive" habit that Peirce identifies with sheer repetition and persistence, attributes that do little to aid in the evaluation of any future event.

Still, Peirce favored abduction because it seemed to be the only office of reasoning that allowed for the arrival of a novel, unprecedented thought. Induction, tied to habit, tends to subsume each event into the Same, and the logical necessity of the deductive syllogism relies on full knowledge of all the premises, knowledge that, by definition, is not available in exploratory scientific enterprise.[14] Thus Peirce sought to describe the persuasive force of abduction—what he sometimes called "hypothesis"—in terms other than those reserved for logic:

> Hypothesis substitutes, for a complicated tangle of predicates attached to one subject, a single conception. Now there is a peculiar sensation belonging to the act of thinking that each of these predicates inheres in the subject. In hypothetic inference this complicated feeling so produced is replaced by a single feeling of greater intensity, that belonging to the act of thinking the hypothetical conclusion . . . We may say, therefore, that hypothesis produces the *sensuous* of thought, and induction the *habitual* element.[15]

Peirce's rhetoric and semiotics sought, among other things, to materialize our understandings of language and conviction, so the "sensuous" character of abductive thinking should not be read through the lens of the Platonic and Aristotelian suspicions of pathos. Instead, Peirce offers it as a description of the operation of the only mode of reasoning that seems fit for an encounter with the future, a deployment of persuasive force that gambles on the unprecedented. Such a gamble works through an intensive practice of substitution—a single concept is swapped for a tangle.

In Christopher Langton's 1987 manifesto for artificial life, precisely such an encounter with the future takes place. The missing term in the abductive transaction is "life." Writing of biology's need to expand its purview and look at material substrates other than carbon, Langton writes that the traditional narrow context of biology(!) makes it impossible for researchers to really "understand" life. "Only when we are able to view *life-as-we-know-it* in the larger context of *life-as-it-could-be* will we really understand the nature of the beast." With this claim, Langton offers a *formalist* definition of life, one that argues that the phenomenon of living systems is tied to organizational and not material attributes. This hypothesis is not itself new—Maturana and Varela's notion of "autopoietic" machines also argues that life is an organizational phenomenon, one that is perhaps independent of its material instantiation. Indeed, the rhetorical practices of "information" that have transformed the life sciences support such a dislocated conception of living systems, as "information" becomes mobile, capable of instantiation in contexts other than its origin. What is peculiar to Langton's abductive move is the claim that one can *only* understand life adequately on the basis of its instantiation elsewhere: "life writ large across all material substrates...is the true subject matter of biology." A larger sample size, so the argument goes, might help us understand both what's peculiar to carbon-based life and to what more general descriptions and laws govern living systems, a kind of metalife.

A tension or even a tangle emerges from this bold move into the "synthetic approach to biology" where organisms emerge *in silico*. What Langton proposes is as novel a shift for biology as the discovery and study of microscopic organisms with Leeuwenhoek's deployment of the microscope. Instead of greater magnification that allows for the representation of a new realm of the organic, the iterative capabilities of the computer make visible the lively and self-organizing capacities of an inorganic stratum. And yet, in its logical formulation, Langton's claim depends upon the very knowledge that it seeks. The very "understanding" that is intended to orient the life sciences—"the true subject matter of biology"—is yet to come, bundled with an analysis of "life-as-it-could-be." In the meantime, our very criteria for identifying and studying living systems remain vague, operational defin-itions haunted by their character as simulacra. The very impetus for artificial life re-search—the lack of sufficient knowledge of the formal attributes of life—stymies what Langton will call the "big claim" of artificial life, a claim that defines *in silico* creatures as "alive."

> The *big* claim is that a properly organized set of artificial primitives carrying out the same functional roles as biomolecules in natural living systems will support a process that will be "alive" in the same way that natural organisms are alive. Artificial Life will therefore be *genuine* life—it will simply be made out of different stuff than the life that has evolved here on Earth.[16]

This claim, anchored as it is in an understanding of the "way that natural organisms are alive," begs the very question that alife allegedly illuminates: What "way," pre-cisely, are natural organisms alive?

But, as Peirce pointed out, the dependence of an abductive argu-ment on a term that is yet to come, off the Earth—such as "life-as-it-could-be"—is not simply lacking in logical coherence; such an opening toward the future allows for the contingent, even improbable arrival of precisely such a missing term. Indeed, Peirce writes of abduction: "Never mind how improbable these suppositions are; everything which happens is infinitely improbable."[17] For Langton, the supposition of a synthetic biology substitutes for, even doubles, yet another supposition:

> Since it is quite unlikely that organisms based on different physical chemistries will present themselves to us for study in the foreseeable future, our only al-ternative is to try to synthesize alternative life-forms ourselves—Artificial Life: life made by man rather than by nature.[18]

How might such organisms "present themselves" to humans? The enjoyment of Langton's joke depends upon our capacity to abide the exchange of one improbability for another: it seems, for some reason, more likely that we can fabricate life than that aliens will arrive any time soon. Along with that other alien contact strategy SETI, however, alife comes up against an operational difficulty: How to determine whether its object is "really" alive?[19]

To this query, Langton implicitly offers an intriguing answer. Discussing the transformation of an artificial life "genotype" or GTYPE into its "phenotype" or PTYPE, Langton traces out an irreducible contingency in artificial life, and perhaps, life:

> It is not possible in the general case to adduce which specific alterations must be made to a GTYPE to effect a desired change in the PTYPE. The problem is that any specific PTYPE trait is, in general, an effect of many nonlinear interactions between the behavioral primitives of the system. Consequently, given an arbitrary proposed change to the PTYPE, it may be impossible to determine by any formal procedure exactly what changes would have to be made to the GTYPE to effect that—and *only* that—change in the PTYPE. It is not a practically computable problem. There is no way to calculate the answer—short of exhaustive search—*even though there may be an answer.*[20]

Assuming that an attribute of PTYPE would be "living," then, no inspection of any GTYPE can yield an understanding of any PTYPE's liveliness. What, then, would provide the alife researcher with an understanding of the liveliness of a PTYPE?

"Trial and error," Langton claims, is "the only way to proceed in the face of such an unpredictability." On this abductive account—it can only make assumptions in the face of the future—alife phenotypes, PTYPES, can only emerge through the actual execution of an algorithm—its translation and "expression," a process that is itself characterized by multiple levels of interaction:

> It should be noted that the PTYPE is a multilevel phenomenon. First, there is the PTYPE associated with each particular instruction—the effect that the instruction has on the entity's behavior when it is expressed. Second, there is the PTYPE associated with each individual entity—its individual behavior within the aggregate. Third, there is the PTYPE associated with the aggregate as a whole.[21]

Thus the actual status of any alife creature cannot be inferred from its initial configuration—GTYPE—and its expression is characterized by levels, thresholds that are themselves the outcome of multiple "non-linear interactions." If life, as Langton claims, is not "stuff," but is instead an "effect," then the effects of liveliness can only be articulated after the execution, as it were, of the alife code.

What I want to suggest is that at each level, acts of translation occur. "Within "the screen, alife organisms survive based on their interactions with both their virtual environment and other alife creatures. So, for example, the strange dogs called "Moofs" that sometimes populate *SimLife* emerge as "translations" of their digital genomes, and one can tinker with the genomes in the hope of tweaking Moof success and behavior, but the ecology even of a simple program like *SimLife* is sufficiently complex that one cannot predict the effect on the Moof phenotype, at least in terms of its behavior within the virtual ecology that it inhabits.

Representing Life for a Living

But once we shift our focus from the alleged interior of the computer, we encounter yet another translation practice. Moofs are inoculated into the virtual ecology of Simlife based on the preferences and habits of the humans interacting with them. The overall success of an alife world—the number of times that a mass of organisms sprout on silicon substrates all over the Earth—depends on their ability to *seduce humans.* That is, their "liveliness"—their ability to achieve the reproductive success and other "lifelike behaviors" in the virtual ecology of the computer—depends on their success in *representing* "life" to their human wetware. This would be simply tautological were it not for the fact that our definitions of life are themselves, at best, recursive—ongoing feedback loops whose origin and destination are quite simply non sequiturs.

Even the definition of life emerges less from a sequence than a tangle, a complex of interaction that conjoins rather than demarcates life from its alleged others, such as machines and avatars. Gaia, for example, both names the emergence of the earth as a superorganism and announces the immense capacities of "matter" to be imbricated and infiltrated with life: It's alive! Lynn Margulis's analyses of symbiogenesis and Gaia both suggest a strange capacity to be invaded characteristic of both the earth and the nucleated cell. In this frame, the eukaryotic cell emerges not only through a parasitic invasion by bacterial DNA on a prokaryotic cell, or, conversely, via the consumption of prey DNA by a predator, but from the supple and specific capacities to be invaded co-evolving with the prokaryotic cell. This capacity to survive an inhabitation is just as important as the bacterial "tolerance for their predators" that allowed select bacteria to live off the hospitality of the cell and avoid consumption.[22]

According to Margulis, it was probably the evolution of a protective membrane that allowed early prokaryotes to survive the "poisonous guests." Thus vital to Margulis's account of the bacterial deterritorialization through which a new form of cell emerged was a form of blockage—a blockage of its DNA.[23] What would be the analogous operation in alife, if there is one?

Perhaps it is in the blockage of the *self* that alife proceeds. As familiars, alife offers itself only through the production of indiscernibility—the sudden hailing of a life-form, a creation of complicity among humans and machines.

This complicity—as with the long-term embrace of mitochondria and the eukaryotic cell—emerges out of a liaison whose very operation entails a temporary blockage or even suspension of the self, a self that would wield control over the machine.

This self must be actively blocked or distracted through visible and sometimes narratable tactics, some of which are more successful than others. Thus the success of alife organisms in their virtual ecology is tied to their success in an actual interactive visual ecology, an ecology also populated by humans, in which, for example, one simply *cannot look away*. The promise of alife is that something is always about to happen...

Hence the cute, perky vitality of most alife organisms. Alife organisms need not attract in this same way, just as all flowering plants do not tempt the wasp as the orchid does.[24] But alife creatures must indeed represent the life effect in a fashion that is visible and articulable with the mass of humans interacting with them. Thus at the level of PTYPE where the speech act "It's Alive!" emerges, alife creatures require a ribotypical apparatus that will render them compelling to their human hosts. If an alife organism falls in the forest, and no one is around to hear it, it only makes a virtual sound. Not because of any allegedly postmodern solipsism attributable to alife creatures, but because actualized alife organisms *represent life for a living*. The alife organisms that achieve the most success in Darwinian terms are those that are most readily and remarkably narrated or otherwise replicated. This is as much an attribute of alife behavior as their feeding habits, and at the level of the actual, it is an obligatory passage point for success. Note that this attribute by definition cannot simply re-present life—what would such a representation look like?—but must instead provoke, seduce it into actualization. As Pierre Lévy, following Deleuze's discussion, writes, "the real *resembles* the possible whereas the actual *responds* to the virtual."[25]

I want to be clear that I am not claiming that alife organisms are simply the result of human "decisions," or that it is *only* the rhetorical softwares bundled with alife organisms that make them lively. Such a humanist understanding of alife would overlook the fact that humans do not simply choose rhetorical practices; rather, they are persuaded by, and respond to, them. The rhetorical softwares of information that transformed biology, for example, were less careful deployments of knowledge than contingent experiments in persuasion. And the success that alife organisms have as *virtual* organisms is not "fake"; it is simply another level of PTYPE, another level of contingency, than the actual.

Philosophers Gilles Deleuze and Félix Guattari articulate this distinction between the actual and the virtual as one based on "chaos," the sheer contingency of an unactualized event, a program that may be successfully run. Real but not actualized, the virtual is a consistency—such as a configuration of code or a spore—that remains to be executed. The virtual is not, therefore, "unreal." Nor does it *lack* actuality—such a description would depend on an abduction of the future, a retroactive understanding of the virtual in terms of its instantiation as actual. Nor is the virtual without the resistances and finitudes we often attribute to the real. It bears its own constraints—the capacity to be rendered into a virtual substrate of code, the materiality of its substrate, and perhaps most strangely, the capability of encountering the difference of the future, to negotiate the catastrophic change in kind that is the movement from the virtual to the actual, program to instantiation, as in Kauffman's "we must instead simply stand back and watch the pageant." Feminist thinker Elizabeth Grosz distinguishes this characteristic of the virtual/actual relation precisely in terms of the occurrence of the future to the virtual, "what befalls it."

> The movement of realisation seems like the concretisation of a preexistent plan or program; by contrast, the movement of actualisation is the opening up of the virtual to what befalls it.[26]

What befalls alife creatures is itself multiple, even a crowd. The evolution and emergence of artificial life forms occurs in relation to both the materiality of alife's virtual ecology—my hard disk is getting full!—and its relationships to other alife creatures. Crucially for alife, though, is the fact that one level of their actualization depends upon their ability to be "befallen" by human wetware, an actual response to the virtual. It should be objected that the privilege of this last level of actualization—the level at which "It's alive!" emerges—is entirely enmeshed with the ecology of humans. Thinking the novelty and specificity of artificial life, though, demands that we encounter the crucial ways in which human corporeality is entwined with alife's status as "life," even if it does not *dominate* alife.[27] In some sense that I will discuss in more detail at some instant in the future, the actualization of alife familiars as "life" is perhaps the least novel level in the alife PTYPE, one that reterritorializes the strange contingencies of the virtual becoming actual into that old saw, "life." Indeed, perhaps the simulation of life is but a ruse, a hoax, or stealth tactic that enables the propagation of entities that dwell much more in alterity than life, novel entities that mime life as a tactic and not an essence.

Artificial life is not, for example, an operation of simple Darwinian artificial selection where humans breed the liveliest alife creatures, consciously or unconsciously; instead, alife creatures' very existence as actualized life, rather than as interesting enterprises in computation, is enmeshed with the human phenotypes with which they interact. The "original" status of alife as life—that which makes it replicable in the first place—is thoroughly bound up with the affect—Peirce's "sensuous character of thought"—of the humans that encounter them.[28] At an actualized level of PTYPE, that level at which alife become replicable as life and spread across the hard disks and networks of the infosphere, the liveliness of alife creatures is contingent on the relations between their effects—such as reproduction—and their ability to be articulated as lively, an ability that does not simply reside in human narrators but is provocative of them.

Still, it is worth considering a general framework for thinking about how such provocation proceeds. Darwin's less heralded mechanism of selection—sexual selection—might connect the representation of life for a living to the work of rendering life. That is, given that "life" and "representation" are different effects whose ecologies do not necessarily overlap, how do alife creatures make a living, representing life for a living?

In both *The Origin of Species* and *The Descent of Man* Darwin reminds us over and over that it is not only through war and struggle that evolutionary fitness emerges. Elaborately choreographed seduction is, for example, an important vector of ornithological sexual selection, or the struggle for mates among birds of the same species:

> All those who have attended to the subject, believe that there is the severest rivalry between the males of many species to attract, by singing, the females. The rock-thrush of Guiana, birds of paradise, and some others, congregate; and successive males display with the most elaborate care, and show off in the best manner, their gorgeous plumage; they likewise perform strange antics before the females, which, standing by as spectators, at last choose the most attractive partner.[29]

While the ascription of choice to the female bird here clearly troubled Darwin with its implications for female agency, he was perhaps more bothered by the sheer exuberance of plumage, seeing in it a monstrosity exceeding any usual notion of fitness.

> The tuft of hair on the breast of the wild turkey-cock cannot be of any use, and it is doubtful whether it can be ornamental in the eyes of the female bird; indeed, had the tuft appeared under domestication, it would have been called a monstrosity.[30]

Darwin's remark here seems to indicate that choice is an inadequate model for thinking these scenes of seduction and their feedback onto ornament, song, and speech. Darwin's doubt concerning the possible excitement of a female bird by the tom's waddle speaks less to Darwin's impoverished ornithological imagination than it does to the weakness of the choice model. Indeed, rather than resulting from a site of careful evaluation and control (such as domestication) the evolution of such spectacularly useless ornaments speaks to the operation not of decision but of emergence: the monstrousness of the tuft sprouts out of a feedback loop whose origin is neither male nor female, passive nor active: interaction. In this sense, alife organisms and humans form an extended, interactive phenotype of each other, with rhetorical softwares serving as the ribotypic translation apparatus that enables this operation of alife code on human bodies and vice versa, the becoming-silicon of flesh, the becoming-flesh of silicon. On Darwin's model, perhaps artificial life is in an instance of trans-species sexual selection.[31]

As with the other levels of translation, one simply cannot tell in advance if a given RTYPE/PTYPE interaction will succeed in yielding organisms that will achieve actualized success as lively. Each rhetorical software or ribotype must be run—and the alife organisms' "lives," as actualized lives, are at stake. It is only in practice—what befalls the GTYPE and is contingent virtual vitality—that the actualized vitality of the alife organism can emerge. Indeed, in some sense it is only in the future that such liveliness can occur, for each practice encounters the news that any formal definition of life is yet to come.

Perhaps there is good reason for the apparently irreducible contingency of the GTYPE /RTYPE /PTYPE interaction. John Von Neumann, polymath propagator of the theory of self-reproducing automata, describes complexity as being more difficult to describe than to practice:

> There is a good deal in formal logics to indicate that the description of the functions of an automaton is simpler than the automaton itself, as long as the automaton is not very complicated, but that when you get to high complications, *the actual object is simpler than the literary description.* (emphasis added)[32]

However one characterizes the life effect, it is certainly "complicated"—it involves passages over borders, the organization of a "complex" or an ecology. As with Kauffman's discussion of irreducibility in Chapter 0, the description or translation of the liveliness of artificial life is perhaps more complex than the automata of alife themselves. As in the joke I mentioned above, it is perhaps simpler to practice or "grow" alife than it is to describe it. Truly an abductive enterprise, alife continually seeks confirmation in a practice that is yet to come—the translation of alife as lively, an understanding of the formal nature of life, life-as-we-know-it within the context of life-as-it-could-be.

More than a logical tangle, though, the tactic of interrogating the comparative probability of artificial life and alien life provokes an experience of life's ubiquity—no longer confined either to the planet or to flesh, it finds itself distributed across the desktop and the universe.

Literary Ribotypes

Not all creatures can be rendered equally visible, narratable and therefore abductable, so some rhetorical tactics would seem to be more successful than others in the evolving ribotype of alife and its operations of transhuman sexual selection. In the literary phylum, authors such as Philip K. Dick and William S. Burroughs have generated remarkable rhetorical effects of vitality in diverse and divergent contexts. Dick, the speed-typing author of over thirty-six novels and several volumes of short stories, describes "vugs," a Titanian, silicon-based life form that inhabits his 1963 novel *Game Players of Titan*.

> They were a silicon-based life form, rather than carbon-based; their cycle was slow, and involved methane rather than oxygen as the metabolic catalyst. And they were bisexual... "Poke it," Bill Calumine said to Jack Blau. With the vug-stick, Jack prodded the jelly-like cytoplasm of the vug. "Go home," he told it sharply.[33]

Dick's figuration of the *in silico* creature quickly overtakes the comfortable distance established between the stick-wielding humans and the amorphous blobs from Titan. Telepathic, vugs render the distinction between the interiority of a human and its "outside" indeterminable, as the boundaries of human identity become as fluid as the physical outline of the vug. Before long, both the characters in and the readers of Dick's novel find themselves "surrounded":

> As he sat on the edge of the bed removing his clothes he found something, a match folder under the lamp by the bed and examined it.... On the match folder, in his own hand penciled words: WE ARE ENTIRELY SURROUNDED BY BUGS RUGS VUGS.[34]

Crucial to the effect garnered here—one might call it panic—is the vehicle of vug knowledge. Writing, that allegedly stable reservoir of memory, becomes the vector not of certainty but of its disturbance. The reception of a message to oneself becomes the occasion not just for recall but for the disturbance of recollection. For Dick's character, Pete Garden, could not remember writing such a note:

> I wonder when I wrote that? In the bar? On the way home? Probably when I first figured it out, when I was talking with Dr. Philipson.[35]

By disturbing both the interiority of his characters—through telepathy, you're thinking—and the interiority of his reader—through the suggestion that Dick's writing, too, may contain a strange message, one so inarticulable that one must try out different bonsonants, ronsonants, consonants—Dick dislocates the vitality of the vug and distributes it across other substrates. The very existence of a silicon life form immediately leaves us, possibly, "surrounded," but Dick's play on the simulatable character of life and the incessant movement of writing implicates both his characters and his readers in a strangely interactive paranoid world where anything, even a book, could be alive.

This dislocation of vitality from its "home," carbon, instills many of Dick's novels with this uncanny sense of being "surrounded" by vitality. In *Radio Free Albemuth*, a novel found with Dick's papers after his 1982 death, Nicholas Brady receives a visit from his future self:

> He had the impression that the figure, himself, had come back from the future, perhaps from a point vastly far ahead, to make sure that he, his prior self, was doing okay at a critical time in his life. The impression was distinct and strong and he could not rid himself of it.[36]

This unforgettable memory of the future seemed primarily concerned with one thing: that the universe is itself alive. Having encountered VALIS—the Vast Active Living Intelligent System—Brady was now "plugged into" the enormous vitality of the "void."

> By now I knew what had happened to me; for reasons I did not understand, I had become plugged into an intergalactic communications network, and I gazed up trying to locate it, although most likely locating it was impossible.[37]

Crucial to Dick's formulation is the notion that such vitality cannot be "located." Beyond the boundaries of any given organism—whether human, vug, or the Universe itself—vitality is characterized by its excess, a surplus that renders the desire to locate the territory of vitality difficult if not impossible. Only violence—the repetitive prodding of the vug stick—connects the flowing turbulence of life to its alleged container, "home," and the network can be "plugged into" but not located, as each node leads to another in the distributed effect Dick renders as VALIS, an artificial, computational but living god.

So too does much of alife create this sense of dislocation, as the incessant vitality, like the Energizer battery bunny, goes "on and on," everywhere. Indeed, as I have noted elsewhere, the practice of alife seems to be involved in the desire for transcendence, the desire to be "above everywhere," a position from which one can articulate, finally, the formal characteristics of life in this moment of its dislocation. But the excessive, transformative character of alife is effected in more than a simply transcendental fashion; the pesky vitality periodically invades the "identity" of the user, that wetware sufficiently charmed by the Tamagotchi to bury it.[38] As a classic installation of subjectivity—the very insemination of the name—nothing would seem more masterful and transcendental than this act of self-replicating memorialization.[39] And yet as a practice of mourning, this electric funeral indexes a complicity with alife as fundamental as any transcendental desire, a vitality at home nowhere and at play everywhere.

Still, as with Dick's novels, such vitality exists within a frame: the virtual ecology of the computer. If, after a time, the reader of Philip K. Dick's novels slowly remembers the memorializing capacities of writing, so too does silicon's vitality remain confined, primarily, to the postvital window through which it emerges. If the operation of abduction—the encounter with a fundamentally alien, unprecedented future—makes the appearance of *in silico* organisms possibly ubiquitous in the distributed network of contemporary "life," what ensures the autonomy and isolation of the vital silicon creature in the context of its leaky legacy of excess?

Life Is for Fetuses; or, Insane in the Membrane

A universe comes into being when space is severed into two. A unity is defined. The description, invention and manipulation of unities is at the base of all scientific inquiry.

> Humberto Maturana and
> Francisco Varela, "Autopoiesis:
> the Organization of the Living"

Life itself, a kind of technoscientific deity, may be what is virtually pregnant.

Donna Haraway
Modest-Witness@Second-Millennium

Humberto Maturana and Francisco Varela, in their 1972 text "Autopoiesis: The Organization of the Living," offer a general theory of living systems that would characterize the specificity of vital systems in terms of their *autonomy*.[40] In this respect, Maturana and Varela are placed firmly within an Aristotelian tradition that saw organisms as wholes mobilized by their purpose or "telos," but with a difference: the "purpose" of an organism is autonomy itself.

This formulation of a theory of biological systems allows one to dispense with the classical category of teleology and its shadow of God, but perhaps more crucially it also enables the evaporation of the distinction between living systems and machines. In the place of this distinction—one based, perhaps, on some vitalist trace in biology—Maturana and Varela generate the difference between "autopoietic" and "allopoietic" machines. Autopoietic machines work on themselves, as it were, generating their identity as an effect of their ongoing self-organization:

> An autopoietic machine is a machine organized (defined as a unity) as a network of processes of production (transformation and destruction) of components that produce the components which: (i) through their interactions and transformations continuously generate and realize the network of processes (relations) that produced them; and (ii) constitute it as a concrete unity in the space in which they (the components) exist by specifying the topological domain of its realization of such a network.[41]

Such a lengthy and tangled definition reminds us of Von Neumann's observation cited above: that an event as complex as a living system is easier to achieve than to describe. Still, Maturana and Varela's definition has serious influence within certain strands of artificial life, so its rhetorical management of this complex problem—the very definition and borders of living systems—is crucial to an understanding of the investment of the interior of autonomous alife organisms with "life."

On the one hand, it would seem obvious that Maturana and Varela's arguments enable artificial life. By decoupling life from its usual lodging (i.e., organisms), autopoiesis makes possible the dislodging of life from any organic location whatsoever. It is in this sense that Maturana and Varela's work resonates with the news that living systems are primarily informatic. They seek to situate the autopoietic effect within a larger integrated *process* rather than confining it to its network of molecular effects, even as they refuse the claim that living systems can be characterized as operations of coding.[42]

This deterritorialization of life via autopoiesis, then, continued the erosion of the machine/organism distinction that was wrought by the ascendance of the postvital understanding of living systems. Unlike the distributed character of networked life, whose vitality is tied to the possibility of a contingent outside with which each component could connect, Maturana and Varela's vision emphasizes the autonomy and closure of the autopoietic system. If the postvital understanding of life emphasizes connection—as in Stuart Kauffman's Boolean nets, where nodes garner vitality and order through relation to multiple, other nodes—then the cybernetic argument of Maturana and Varela renders life as an interiority, one constantly making itself *as a self.*

This notion of the *interiority* that inheres in autopoietic systems maps logically onto Maturana and Varela's claim that autopoiesis is "necessary and sufficient for the occurrence of all biological phenomena."[43] This necessary and sufficient status accorded autopoiesis retains the historical sovereignty and interiority of life—the agency of an organism consists in its construction only of itself—even as Maturana and Varela seek to offer a theory of living *organizations*, a theory that stresses the relentlessly relational character of living systems.

This self-contained logical character of the autopoietic system—as necessary and sufficient—marks the topological map of the living system as well. "In the beginning," to paraphrase the quote with which I began this segment, "was the inside and the outside." The authors begin with this distinction, warranted by their emphasis on "autonomy," what they deem to be "so obviously an essential feature of living systems."[44] For a distinction that possesses so much self-evidence, the claim for the autonomy of the autopoietic system—that process of self-organization that emerges between the inside and the outside—poses many problems for Maturana and Varela. Even as they attempt to demarcate the distinctive qualities of living systems, they find themselves unable to either confirm or deny the difference between social organizations and biological ones. Faced with what they see as the ethical problems that inhere in the answer to such a question, problems they deem to be the problems of the future, Maturana and Varela defer the answer to this question to the future itself:

> In fact no position or view that has any relevance in the domain of human relations can be deemed free from ethical and political implications ... This

responsibility we are ready to take, yet since we—Maturana and Varela—do not fully agree on an answer to the question ... we have decided to postpone this discussion.[45]

That is, Maturana and Varela cannot agree on the status of the following question: Are social organizations inside or outside the purview of biological laws? The decision not to decide, to postpone or defer, allegorizes the futural character of the very membrane between inside and outside, the autonomy machine, that Maturana and Varela deploy. Only retroactively—i.e., in the future—is the distinction between inside and outside self-evident. Far from obvious, the Möbius space of inside and outside are sites of indeterminacy and undecidability that *emerge* in the abductive process of living systems. The very autonomy of them—Maturana and Varela—is threatened by the force of the problem: they cannot choose not to decide about the future, ethical problems posed by the theory of autopoiesis they offer.[46] Even on Maturana and Varela's own terms, such a topological distinction poses a cognitive problem that "has to do with the capacity of the observer to recognize the relations that define the system as a unity, and with his capacity to distinguish the boundaries which delimit the unity in the space in which it is realized."[47]

Despite its conceptual trouble, the localization of life rendered by the theory of autopoiesis does much to ensure the confinement of artificial life in its window. For with its clear, if tangled, exposition of the claim that autonomy is the fundamental life "behavior," they buttress the continuation of the classical claim for the interiority of organisms even as the life effect is distributed across the multiple links of networks.

By propagating such a clear demarcation between the inside and outside of the living system, Maturana and Varela's argument replicates the historical comportment of life as an entity distinguishable from its environment. The contemporary localization of life extends this encapsulation and treats the fetus as an entity distinct from its mother's body. As scholars such as Susan Squier, Barbara Duden, Karen Newman, Valerie Hartouni, and Donna Haraway have argued, the fetus was "born" as a distinct entity through rhetorical and visual techniques that severed it from the maternal body and invested it with "life" and subjectivity. Both the rhetorics of "choice" and "prolife," Newman argues, emerge out of a discourse full of rights-laden bodies, individuals in direct conflict that paradoxically dwell in the same body. While Newman overlooks the historical transformations of "life" in the period that she analyzes, she carefully documents the persistent, historical occlusion of the maternal body and the emergence of the fetus, an emergence that functions through the attribution of distinct interiority to an entity that is, paradoxically, *inside* the invisible maternal body.

Donna Haraway, in "The Virtual Speculum in the New World Order," highlights the particularly odd status of such a distinct fetus at a moment when life has been dislocated. Haraway offers multiple readings of a cartoon that she dubs "Virtual Speculum," which features

> a female nude . . . in the position of Adam, whose hand is extended to the creative interface with not God the Father but a keyboard for a computer whose display screen shows the global digital fetus in its amniotic sac.[48]

Among Haraway's proliferating readings, she argues that the digital fetus is "literally . . . somehow in the computer" and thus "more connected to downloading than birth or abortion . . . the on-screen fetus is an artificial life form."[49]

In Haraway's formulation of the "fetus *in* cyberspace," the topological comportment of life as outlined by Maturana and Varela returns; life is conceptually or "virtually pregnant"—disturbed at its border between "inside" and "outside," a fetus "in" a nonspace, life has missed its (historical) period. The conundrum posed by *Virtual Speculum* is literally: Where is life? Its "source" appears to be the gleaming screen of the workstation, as capital transforms more than the global markets via the new technologies of pixel, keyboard, and perhaps network. But such a reading immediately overlooks the corporeal connection of the female nude, whose hand touches the keypad, whose digits, perhaps, experience the pain of labor via carpal tunnel syndrome. Thus the "digital" fetus, awash in amniotic and semiotic fluid, exemplifies the persistent exteriority of living systems. Even under its greatest ideological pressure to be an interiority—the fetus in a box—life resists attributions of autonomy as effectively as genetically modified corn finds its way into a taco shell.[50] Impossible to locate, the life effect occupies a Möbius body, a rhizome that traverses the interiority of the screen and its outside. As a membrane, the screen marks less a clean boundary than a multiplicity. Difficult to narrate—which is before, which after?—such a multiplicity fosters the implosion of the virtual and the actual even as it highlights an odd morphology of life whose representation is a fluctuating network rather than an organism.

The sheer strangeness of this entanglement—the becoming flesh of silicon, the becoming silicon of flesh—seems to foster an abduction, one that *forestalls* the difference of the future by substituting "for a complicated tangle of predicates attached to one subject, a single conception," i.e. life. Newman describes this demand as a "referential panic, a need for realist images" that would render the strange new tangles of technoscience and "life" consumable and narratable.

And yet this panic is not confined to the work of representation—alife's power emerges not out of the barrel of a gun, but from the gestures of mouse and pixel, signifying and asignifying grapples with the machinic phylum. To be sure, the familiar seductions of alife are rhetorical, but they involve the sculpted, implicitly choreographed movements of bodies as well as the affects provoked by the encounter with alife creatures.

T   W   o

Simflesh, Simbones:
At Play in the
Artificial Life Ribotype

(10) Enter

AS BOTH familiar and attribute of a network, life has been distributed out of the organism and into the living room. Formerly attributed to the cyborg assemblage of organism and machine, announced by the tell tale metronomic, metonymic, cybernetic beep and the flashing of a pixel on its screen, vital signs now refer to an intense textuality at play in the life sciences, a textuality in which "life" chiasmatically implodes into "information," and signs ecstatically code and decode themselves to vitality in an out-of-body experience. No longer distributed among the machine/organism nexus, vital signs now refer to the sovereign and self-referential status of the timeless immortality of deoxyribonucleic acid. Between DNA and its effects, we have the perfect geometry of an arrow, a diagrammatic vector that at once marks out and erases the emerging deterritorialization of organisms.

As I have argued in the first segments of this book, this intense distribution and outsourcing of life is more than a shift in scientific perception—it alters the contours of corporeal experience, as life ceases to be confined to the interiority of a body and becomes capable of inhabiting locations between bodies: networks, futures, virtualities. In this segment I want to map out the peculiar and sometimes spectacular becomings enabled by one such body practice: *SimLife*, a popular and very trailing edge artificial-life program by the Maxis Corporation. As recipes for becoming, *SimLife*'s algorithms of vitality create quirky linkages between interactive entertainment (inter-tainment) and technoscientific articulations of life. As interactive, affective familiars that emerge only out of that ecology of response I sketched in the first pages of this book, these artificial life-forms hack intimately and delicately on the tortuous and twisted relations among humans and machines. As life becomes a distributed response that emerges only through a fluctuating encounter of multiplicities, the always troubled distinction between "nature" and "culture" becomes less an opposition than an agonistic and, as I invoked above, *sexual* interface. Cryonics—the organized freezing of animal and often human remains—emerges out of a similarly distributing response: vitality becomes distributed over time as well as space. Indeed, in some sense both artificial life and cryonics *are* generative and exuberant responses to the new connectivities proffered by the post vital life sciences. Artificial life, of course, emerges from our arrow: the conflative connection of life and information makes it possible. In that light, cryonics might be seen as an odd vestige of the old corporeality, where the body, like the buggy whip, persists long after it is "needed." Such a judgment, though, forgets the retooled nature of the post vital body; it is not lost or forgotten so much as *in transit*, becoming code—the cryonic body is hooked up to the future, a future I take up in later pages.

In mappings that follow, I hope to highlight the possibilities for lines of flight, political, ecstatic and otherwise, in the emerging nexus of bodies and machines that I will outline here. Stolen from Gilles Deleuze and Félix Guattari, "lines of flight" emerge with effects that are, as we used to say, *something else*, blockages of contemporary formations of power/knowledge.[1] For although the perfect geometry of the arrow, the informational vector that, for much of hegemonic molecular biology, leads directly from DNA to "us" without pausing for difference, we must not forget the possibility that the arrow makes possible a movement elsewhere.

(20) MOUSE EVENT: Double Click on "Avatar"

It's in the laboratories where genetic fantasies are the last avatar of the western dream, the ultimate point of this double and dangerous ambition: purity of choice and descendence.

Jacques Attali

He is not seeing real people, of course. This is all a part of the moving illustration drawn by his computer according to specifications coming down the fibre optic cable. The people are pieces of software called avatars. They are the audiovisual bodies that people use to communicate with each other in the Metaverse.

Neal Stephenson, *Snowcrash*

As an "avatar" of the Western dream, genetic engineering and its attendant fantasies are not merely the natural outcome of a scientific and cultural polymerase chain reaction that began sometime in the 1970s, although that would be a fine metaphor to pursue.[2] Instead, biotechnology and its signs stitch together an avatar, a virtual site of interaction where nucleic acids, science fictions, softwares, the New York Times, and some persistent dreams of the West flash on and off in a complex morphology of the "gene" or "DNA." This morphology itself is a dynamic one, an interactive, animated narrative of technoscience and culture that, like the molecules it describes and inscribes, must be seen to be complexly and rhizomatically in play. Here the distinctions between life and information, lab and living room, segment and rant, scientific and popular culture implode, collide in a space "beyond" the computer screen. No longer pressing our collective noses up against the glass or mirror of nature, we now find that we have passed through to another side, one where the oppositions between nature and culture confuse, collide and ricochet in a unpredictable matrix of work/play/simulation. Scientist George Wald, as quoted by J. D. Bernal in 1967, sums up an isomorphic crossover, one triggered by research into the origin of life:

> We have been told so often and on such tremendous authority as to seem to put it beyond question, that the essence of things must remain forever hidden from us; that we must stand forever outside nature, like children with their noses pressed against the glass, able to look in, but unable to enter. This concept of our origins encourages another view of the matter. We are not looking into the universe from the outside. We are looking at it from inside. Its history is our history; its stuff, our stuff. From that realization we can take some assurance that what we see is real.[3]

This movement from the "outside" to the "inside" is instructive. Whereas much traditional scientific subjectivity describes a scientific gaze as cleft from its object, everywhere and nowhere, hovering above the earth, Wald articulates a complicit empiricism, one that finds "assurance" in the irreducibly situated character of scientific knowledge production. On one register, it is this very solace that needs to be marked here, the metaphysical comfort derived from the proprietary identification of "its stuff" as "our stuff." In this move of "assurance," the world is once again rendered as a mere resource for humans, "us," collaborating in the tedious repetition of transcendence, one that finds certitude in seeing humans "on" the earth.

At the same time, Wald's move—the transformation of the human/nature opposition into a membrane of complicity—names the strangely ecstatic operation of "looking at it from inside." "Its history is our history" disables any hygienic corridor between "nature" and "culture" and suggests a complex of entanglement in which humans are both "on" and "inside" the earth and cosmos, chaosmosis. The infinite proximity of history provokes an extraordinary and not simply semantic exposure to the universe, an exposure "we" now suffer differentially and enjoy. This "realization" is not necessarily mediated primarily by a patient gathering of knowledge but by an ecstasis, the sudden surprise at the proximity and yet sheer alterity of the universe. Implicated and entangled in "stuff," Wald's gaze operates as an excited attractor for becoming as it provokes the experience of the incessant and, not necessarily computable interconnections of "stuff."[4]

This recipe for becoming—a disciplined practice of transforming the relations between human interiority and its diverse outsides—is cultivated in the new fluidities of inside and outside *rather than their disappearance*. The turbulent practice connecting inside and outside in Wald's analysis provokes not simply because of a logical or epistemological disturbance, but through an almost tactile experience of connectivity. This connection forged of human and universe risks yet another campy colonization—It's ours!—but it also marks an ecstatic encounter as humans become extrusions of, rather than observers on, the universe. As with Burroughs's treatment of becoming-gun, Wald's transformation of the border proceeds when a certain blockage is blocked—that blockage called "autonomy."

Deleuze and Guattari's notions of the "Plane of Transcendence" and the "Plane of Consistency" help to map this instability—rather than undecidability—of inside and outside. "The Plane of Consistency" maps the capacities for difference—Kauffman's persistently rugged landscapes—that enable this qualitatively different movement from the outside that Deleuze and Guattari characterize as a movement of immanence. Here:

> There is no structure, any more than there is genesis. There are only relations of movement and rest, speed and slowness between unformed elements, or at least between elements that are relatively unformed, molecules and particles of all kinds.... Nothing subjectifies, but haecceities form according to compositions of nonsubjectified powers or affects.... We therefore call it a plane of Nature, although nature has nothing to do with it, since on this plane there is no distinction between the natural and the artificial.[5]

Deleuze and Guattari's treatment of immanence highlights the relation between the natural and human world as one of "composition," a diverse mixture of speeds and matter that refuse the organizational binary of artificial/natural; the plane of consistency is a massively connected network through which the artificial and the natural emerge. Rather than an axis that demarcates the world into the nature/culture or human/inhuman dialectic, the plane of consistency maps the irreducible particularity ("haecceities") of that strange mixture of qualitative difference, the world. As such, the plane of consistency maps more of a "polyverse" than a universe, one composed of capacities for difference more than unities or things.[6]

Composed of distributed bundles of difference, the plane of consistency is not simply "our stuff." Propriety is an artifact not of immanence and consistency, but of transcendence and organization. Thus the subjectless yet individual space of the plane of consistency is constantly in danger of a sudden nonlinear crash into organization:

> The plane of organization is constantly working away at the plane of consistency, always trying to plug the lines of flight, stop or interrupt the movements of deterritorialization, weigh them down, restratify them, reconstitute forms and subjects in a dimension of depth.[7]

It is within this double formation—an encounter with an immanence prior to and constitutive of subjectivity that is continually and catastrophically stratified by transcendence—that I wish to map the "outside/inside" movement announced by Wald. For although "stuff," in all its particularity and alterity, is ultimately put under arrest on the plane of transcendence via the sign of "our" and its network of propriety and property practices, Wald's description serves as more than an occasion for a deconstruction of the commodification of the world. It offers an encounter with the ecstatic *complicity* between the individuated forms of "us" and "stuff." It renders a dynamic holism that is not an organicism, a massive network of differential connection and becoming that transforms the Earth into something other than a battleground for nature/culture. It resounds with "us," it responds to "stuff."

(25) If Rat Work, Then Line of Flight

On the contemporary life sciences channel, where biology's new reagent is information, Wald's glass, which serves as a quasi plane of consistency that takes in the organic and the artificial, the "universe" and "us," becomes a screen. Artificial life, the synthesis of living organisms in software and robotics, marks a potent and uncertain node in this networked reshuffling of the organic and the inorganic, real and artificial. By looking to the computer screen as a vantage point from which to construct a "universal biology"—not merely life as it is, but Langton's "life as it could be"—artificial life indexes the seismic rhetorical shifts underway in the contemporary life sciences.[8] It *also* maps the intermittent movement of Life, Nature, and Co. to the playful but delimitable space of screens, a movement of location that reinscribes the transcendental valence of "life" and "nature" even as our simulation practices shatter, through an incessant doubling, the priority and sovereignty of vitality and nature. This recoronation of life and nature works precisely to the extent that technoscientific practices occlude their ecologies of possibility, other practices on all sides of the screen.[9]

But if it is sometimes difficult to tell which side of the screen technoscientific practices are on, it is equally difficult to extricate artificial life techniques from lay and popular discourses that suffuse them. Networked with more than other computers, connected to more than disk drives and Web sites, the screen that so uncertainly conjoins and divides the organic and the inorganic is itself a mere polyp in the reef of material strategies that makes possible this new distribution of vitality. While not contesting the claims of artificial life in the name of the natural or the organic, I do want to introduce a flicker of hesitation into the discourse of transcendentality that accompanies and often enables artificial life, a site where the screen becomes a plane of organization and transcendence, hovering over the world of actuality, defining if not determining our knowledge of "life." Computer Scientist and Unabomber target David Gelernter's notion of the computer screen as "Topsite" in *Mirror Worlds* typifies the rhetorical operation that converts the screen into a plane of transcendence: "Topsight is what comes from a far-overhead vantage point, from a bird's eye view that reveals the whole—the big picture; how the parts fit together."[10] Topsite is the spatial version of technoscientific monotheism, its plane of transcendence; if only we could achieve that one perspective with which to view the "whole." This intense intersection of surveillance and knowledge finds its plausibility within a world of implosion described by Gelernter as

> an event that will happen someday soon: You will look into a computer screen and see reality. . . . When you switch one on, you turn the world (like an old sweater) inside out. You stuff the huge multi-institutional ratwork that encompasses you into a genie bottle on your desk. You can see over, under and through it. You can see deeply into it. A bottled institution cannot intimidate, confound or ignore its members; they dominate it. . . . People will stop looking at their computer screens and start gazing into them.[11]

Again, the world becomes "our stuff." Anticipated as a static space of domination, the connections of "world" and "us" operate only in one direction. But such a topological reduction, wherein human beings are somehow exterior to the objects of their gaze, is clearly impossible: one need not be a Derridean to agree that such a quest for spatial-temporal closure is not just unlikely, but hilariously and vertiginously impossible. For obviously, "reality" includes, but is not limited to, our interaction with it, and Gelernter's masterful gaze would need to include a small window that featured the back of his head in front of the screen, and inside that window we would need another that featured the back of the device that rendered his head, and so on. This is not simply a philosophical quandary; the networked movement of the real to the screen has profoundly altered our practices of war, our notions of medicine, and our theoretical understanding of life, forming a tangled feedback loop with the object that is allegedly "contained" within the screen: wetwares. Perhaps most crucially, this inside out movement takes place "sometime soon" in an uncertain practice of anticipation, a rhetorical enterprise I will return to approximately fifty-six percent of the way through this text. For such an anticipatory gaze is not simply passive, as in Kauffman's agonizingly gorgeous waiting game above. Anticipation marks the complicity of Gelernter's vision with an unlikely and thoroughly aleatory welcome, a hospitality to the future which is not simply masterful but receptive to, solicitous of, difference: Welcome to Topsite.

      If the timing of such a welcome is blurred into and by an uncertain future, so too is the very location of this inside/out arrival in doubt. "Ratworks" are composed precisely of connection—exits, entrances, traversal, and flight. Folded "into" the space of the computer, such ratworks transform the relatively locatable sovereignty of the box into the distributed connectivity of a network. Such a connective space tends to cultivate the rugged, contagious networks Kauffman has researched, ecologies of great sensitivity and tremendous "capacity" for transformation.[12]

      Thus I will argue for the screen otherwise: Conceived as a space of becoming that suffuses humans as much as it "contains" the world, the computer screen and its beyond becomes some other topology, an ecstatic multiplex of fluctuation. Neither inside nor outside the computer network or its other, "the world," this wetware is the material, corporeal skin of contemporary transformations of power-knowledge, a material and biological membrane that *is* contemporary capital and its work of deterritorialization. As transformations rather than objects, such flickering assemblages of flesh and money are accessed through what I cited earlier in this text as *itineration* or following. As with the ratwork of markets with which they are so complexly enmeshed, such webs cultivate less mastery than response, and they allow no "fixed point of *view* that is external to what is reproduced."[13] "Rather than external perspectives on a flow, itineration breeds new forms of complicity in which perspective and identity are the very media of response—*they are at stake in the observation.*

This complicity emerges partially through the capacity of virtual technologies—which are neither here nor there—to distribute corporeal practices across a network. Mathematician and semiotician Brian Rotman has pointed out that the mouse, as an interfacial familiar, provides us with a metonym for the project of rendering the body, virtually.[14] The mouse is the becoming-virtual of the finger, the finger's avatar on the other side of the screen. As such it can be described within the McLuhanite rhetorics of prosthesis: the mouse is an extension of the finger "into" electronic space. And yet that extension also troubles the status of the finger, dislocating it, citationally distributing it both inside and outside of the computer screen in a becoming-mouse or even animal of the hand.

Another mouse allows us to flesh out an articulation of this ec-static body. BioMedic Data Systems[15] announced the arrival of the Möbius mouse in a 1992 advertisement for its Electronic Laboratory Animal Monitoring System (ELAMS™). "We keep lab animals from having an identify complex" could be read in at least two ways. On the one hand, it announces the practices of precision and exactitude that compose the contemporary life sciences, practices that are in this in-stance techniques of surveillance. Thus the ability of ELAMS™ to prevent "an iden-tity complex" testifies to the massive disciplining of the mouse that goes on in the lab: ELAMS™ allows the user to know where and how any particular mouse is, "who" it is, in some sense. There will be no doubt about who this mouse is; the ELAMS™ allows the mouse to become code and yet remain distinguishable—genet-ically identical, the "mouse of tomorrow" still needs an identity if it is to play its cru-cial role in knowledge production.

> It can link any animal to any computer database, allowing you to individualize your animal using your study number. You can even characterize them with clinical observation codes . . . Simply put, it replaces the complexities and in-accuracies of toe clipping, ear tagging and tattooing with a foolproof, fast and economical method of positive identification.[16]

Thus the ELAMS™ replaces the practices of the body—surveillance tactics of the toe, ear, and skin—with the operations of coding. The mouse as an object of study is "characterize(d)," transformed into a character, a diagram, an inscription. As such, our mouse exemplifies N. Katherine Hayles's description of the posthuman as the informatic pattern of the mouse becomes more interesting than its material instantiation.[17]

And yet the visuals of the BioMedic ad provoke another response. The mouse/computer relation is rendered as a deterritorialization, a movement that traverses both sides of the screen, beginning and ending everywhere and nowhere. More than a clean deletion or replacement of the body, the screen marks a vertiginous blur of recursion—the becoming code of the mouse and its subsequent territorialization onto identity and deterritorialization "into" citable and thus virtual codes. As such it is not pure code (no implosion of the mouse and code is complete) but is instead a bramble of feedback between a deterritorialization that allows for a reception of the mouse at a distance—the telecommunicative aspects of the BioMedic identity chip, the emergence of the genetically identical "mouse of tomorrow"—and the territorialization of the mouse into a cluster of knowledge that constrains and renders visible the massively engineered critter as a more intensely scientific object. The mouse body thus rendered yields a complex, an assemblage of corporeality and virtuality that traverses both and yet is nowhere in particular, yielding information that is itself captured into formats and frameworks of knowledge production. The ad instantiates a wetware over time, frozen into a single frame—on both sides of the screen at once, the mouse images a spatial and material organization of the machine/body relation as a distribution, an experimental rendering of the unplaceable chaos of actualization that is the production of scientific knowledge and its generation of value.

This sense of the incessant and intense relations that compose the mouse as a scientific object disturbs the easy transcendental position of the human (in this case, the reader) placed in front the screen. For if the mouse traverses both "sides" of the screen, dynamically embodies a Möbius strip of an interface, then humans get dislocated from Topsite, a simple looking down on the mouse, and implicated in the practice of becoming code, wetwares enmeshed with the ubiquitous processes of encoding—the capture of environments by regimes of signs. These codings are then captured by the production of knowledges that sculpt and anticipate the very future of homo sapiens and biological systems in general, a future itself essentially unpredictable and by definition not yet actualized or instantiated.

It is clear that the sense of complicity such a reading fosters is itself disciplined out of most contemporary representations of technoscience; technoscience is, among other events, an autonomy machine that preserves its boundaries. The transcendental position of the scientist is constantly reinscribed in a retroactive production of a recursive declaration: "Identification is who we are," announces

BioMedic Data Systems, as if that were an answer.[18] Hence my intervention here: Keeping the plane of consistency connective, in flight, I want to highlight the rhetorical practices that enable transcendence and subjectivity to sprout from the corporeal and rhetorical mixture that makes up the recursively Möbius-like phenotype and ribotype of alife. In response to such moves I offer precisely a response, a rhetorical encounter of the lab as well as the living room: *Get into* the Sim.

(29) x = year

(30) For x = 1953, Next X

(31) x = x + 1

(35) If x = 1987, then 40

(40) Print "<u>Greetings Electronic Biologist!</u>"

To exemplify the ways that the "gene" has been figured within the new complicities of "life" and "information," "nature" and "culture," I double click first on a software avatar called *SimLife: The Genetic Playground. SimLife*, an old software package available for both PC and Macintosh platforms, provides us with an artifact and a toy that we can treat as a diagram of the complex and interactive inscriptions that suffuse the "gene" in popular and scientific culture. Released by the Maxis Corporation in 1992, *SimLife* is an artificial life program, a "game, a toy and an experimental tool to learn about life, real and artificial." The package consists of a registration card (cited above), a manual, a lab book for recording your SimLife experiments, and a couple of disks that go into the computer. It's no longer supported or marketed by Maxis, but I bought it at a garage sale, doing my bit for biodiversity.

This extraordinary program offers users the ability to be a virtual genetic engineer, as *SimLife* provides an environment for the cultivation, nourishment and breeding of artificial organisms, something akin to cybernetic sea monkeys. Here I will offer a brief analysis of both the documentation and the interactions available to the electronic biologist, the one who has realized along with Harvard molecular biologist Walter Gilbert that biology's new reagent is information.[19]

Already, the hard copy documentation bundled with *SimLife* announces recipes for becoming: Rhino-headed tigers, ostrich giraffes, and a toucan lizard crowd the environment of the cover, eyeballing the reader in a reversal of the gaze to come, a gaming gaze that will look into the screen of a computer and not at it. With the cover, *SimLife* has already begun its tutorial: look into our chimerical eyes. *SimLife*, after all, is itself a chimerical beast—part toy, part game, part tool, it exists at that node where the practices of science and the practices of culture collide.

The *SimLife* manual, as well as the program, goes to great lengths to resist the appellation of mere game:

> SimLife isn't exactly a game—it's what we call a Software Toy. Toys, by defi-
> nition, are more flexible and open-ended than games. . . . In SimLife, the "toy"
> is a biology laboratory in a computer.[20]

While little may be seen to be at stake here in a genre distinction—what matter if we are dealing here with a toy, a game, or a weapon?—the emergence of simulations as "flexible" forms of inter-tainment and inquiry marks not only the blurring of distinction between lab and living room and the subsequent distribution of knowledge production (e.g. Linux); it also highlights the fundamentally experimental and itinerative character of the SimLife experience. As unpredictable and contingent instantiations of simple rules, SimLife creatures promise to surprise the attentive and caring electronic biologist. This promise of affect—*"What? The Moof's have died off?!"*—indexes the ways in which SimLife creatures emerge out of an affective as well as silicon substrate even as the laboratory follows our mouse into the computer. That is, as I argued in the previous episode of alife analysis, the open-ended character of alife creatures—their exposure to the difference of the future—also links them affectively to those humans who would play with them. It is in this sense that the becoming code of vitality selects for, and is grafted to, the responses of its human and machine interactors: There are affective protocols as constitutive of SimLife as those of MS-DOS, Windows, and Macintosh, the operating systems with which they must be run. These affective components of computational environments can be variable—frustration, surprise, and especially laughter emerges—but in all cases such response is an element *of* the interface rather than its simple result. Response, in the case of SimLife, is incorporated into the ecology to the extent that it operates as a vector of sexual selection, a provocation into response: More Moofs, please.[21]

As I argued earlier, this response is most easily understood in terms of the tactics of Darwinian sexual selection, where the "antics" of a male bird and his plumage seductively renders female reproductive agency through intense interaction. In our instance though, one can be even more precise: What is being "reproduced" here are not simply creatures but affective states whose effects are contagious enough to result in a very specific itinerary: that of the mouse. Brian Rotman has produced what he calls "mousegrams," the visualization of those highly choreographed but immensely contingent movements of computer input devices at play in, with, the interface. This responsiveness is not merely the dance of the Moof for its human observer—with a click, wetwares too join the tangle.

(45) MOUSE EVENT: Double Click, Watsonia

What is it that SimLife simulates? How does it inter-tain? Click Click to find out, on the SimLife world I will call "Watsonia" in honor of James Watson's wonderfully SimLife ethos noted earlier: "In order to know what life is, we must know how genes act."

You have already learned two of the technologies of the gene associated with SimLife: look *into* the computer, and when in doubt, double click. Given the constraints of the textual realm I am operating in here, I can only simulate the click, but you'll get the picture. Follow along as I narrate a sample world creation in SimLife.

After entering the SimLife environment through a strangely effective and yet contradictory performative, as I really do press down the following keys—SIM—we begin the first part of the game/simulation: a genesis fantasy. No longer the province of God's finger, genesis has been democratized to include anyone with a mouse. Clicking on "experimental," we get to choose our own creatures, control the creation of our own world, set its climate, watch it run from that space Gelernter refers to as "Topsite." *SimLife* is a transcendental toy; it allows the look into the computer screen to be a look down, as the first sight of the *SimLife* screen is a surveillance shot; lest we forget that *SimLife* was developed with the help of Los Alamos and its new-age complexity franchise, the Santa Fe Institute, we can see the military gaze built right into our game/toy/tool. Looking down and around our world, we can take it all in, without any night-vision goggles.

I have already done some creation work, and the first gaze of Watsonia features the frenzied movement of crawling sea urchins, sprouting sea lettuce, some annoying, chimerical dogs called Moofs, and my personal favorites, those persistent trilobites. While the fantasies of control this game/tool/toy makes possible are obvious—fantasies that Watsonia might very well share with Watson—a perhaps less obvious effect of SimLife is that other fantastic operation: being-out-of-control. With the complex interactions of climate, speciation and mutagens, even the advanced *SimLife* player/researcher/creature will quickly get that exhilarating, positive feeling of a lack of control; the sea lettuce has mutated to swim, the trilobites have learned to fly, and there are no male Moofs left. How did that happen? When in doubt, double click.

Click on History, and you'll find a fine map of SimHistory. History here is a list, a "running record of all events in the world." No background matrix, no narrative, the world of Watsonia is chronicled as a series of discrete, and often catastrophic, events. Interactions take place in the background; despite *SimLife*'s goal that the user "understand that the real world with its millions of species with their combined billions of genes are interrelated and carefully balanced in the food chain and the web of link" or perhaps because of it, History is a list of extinctions, sproutings, and mutations. There is no web of relations connecting these events. The SimNarrative emerges through the clicking, reading, fluctuating wetware of the user.

History's status as a sequence of discrete, static events instantiates the view of life expounded by Watson above: Life exists in the gene, genes are the actors of life, and bodies mere supplements or extensions of the miraculous software called DNA. Indeed, in SimLife, bodies are transparent: If you want to see the genome of a SimLife creature, then click on it with the double helix, and a window zooms out to exhibit the control panel/genome that determines the characteristics of any organism. And yet, according to the *SimLife* manual, this is not what these organisms *really* look like:

> As you play SimLife, the different plants and animals will visually appear in a few different ways. None of these ways truly and accurately shows the way these organisms look. These electronic organisms exist as ones and zeros—energy states in transistor switches in the memory chips of your computer. Assuming that most of the beings that play SimLife are human, and that none of the humans we know can see energy states in transistor switches, we figured we'd better find some way to visually present SimLife-forms in a way that humans can see and understand.[22]

This nod toward visual representation—an ornamental representation oriented less to survival than to seduction and replication—points to the importance of Barbieri's claims concerning ribotype and to the rhetorical functions of such ribotype, rhetorics that allow the unprecedented *in silico* organisms of SimLife to be seen and "understood."

Thus the implosion between "real" organisms and "SimLife" organisms is complete, leading to a reversal; no longer what we see on the computer screen, SimLife creatures now dwell *inside* the computer, beyond the screen. And yet these ones and zeros are both the "organisms" and their "genes"; as in a certain version of molecular biology, organisms are nothing but (electronic) DNA. In SimLife, what is "real" about an organism are its genes, code that dwells in the computer. Simlife "bodies" are energy states about to fluctuate. Whereas the chimerical cover and genre of *SimLife* announce strange mixtures of "phenotypes," the location of the "real" appearance of SimLife creatures as inside the computer—*You stuff the huge multi-institutional ratwork that encompasses you into a genie bottle on your desk*—maps out the overtaking of phenotype by genotype that has characterized the ascent of genomics. The pixels of *SimLife* become blurred reflections of the real SimLife organisms, organisms that dwell not on the screen but in it, through a glass darkly.

While we may *still* press our noses up against the screen, looking in vain for the true and accurate picture of the real, we nonetheless have access: clicking on the mouse gives us an avatar, a simulated body of a creature under the purview of "Charles Darwin (only in his dreams),"[23] one of the suggested subject positions for the *SimLife* user. What it gives us access to is another question, one I can only speculate on, since I have not yet played *SimLife* long enough. It plugs us into the new truth of bodies as extensions of our genetic software, our new understanding of "what life is." As model rocketry in the heyday of NASA enabled many to participate in that "giant step for mankind,"—what William S. Burroughs has called getting "the whole show out of the barnyard and into Space"[24]—electronic biology plugs us into, makes us virtual witnesses of, genomics. It connects us to the circuits of culture and technoscience that bring biotechnology to Wall Street, Washington, and finally, our SimLiving rooms.

Of course, it is the status of this circuit that I want to question. Whereas "life" is ascribed to the autonomous and isolated little bits of energy occupying the computer, I want to highlight the importance of what I above characterized as the ribotype in the constitution of SimLife. For it is with rhetorical operations that the vitality of SimLife organisms is located on the screen, indeed *located* at all. While philosophers may grapple with the metaphysical question—Is alife *really* alive, or not?—I want to highlight the possibility conditions that allow this to be posited as a question at all. For what makes SimLife "lively"?

I have thus far focused on the interactions between *SimLife* and its users for good reason: against the grain of the claim that alife creatures "themselves" enjoy vitality, I want to suggest that the screen that orients alife is on a plane of immanence, a massive assemblage of machines, users, and rhetorics that semiotically and materially distribute the "vitality effect." The relation between the user and the Simlife organisms is more one of ongoing, affective interaction than demiurgical creation, a theological subject position that SimLife, as well as alife in general, constantly alludes to.[25] And this ongoing interaction produces effects on the user as well as the SimLife organisms. Karl Sigmund, writing in *Games of Life: Explorations in Ecology, Evolution, and Behaviour*, remembers the need that Conway's game of Life—an early precursor to contemporary alife—had for human wetware:

> In the early 1970s, at a time when computer viruses were not yet an all too common plague, there was another type of epidemic causing alarm among computer owners. It [Life] used the human brain as intermediate host.[26]

While the status of Sigmund's description here is uncertain—*is he serious?*—the recognition of the material substrate necessary for Life to propagate is well put. Neither masterful creator nor objective observer, the player of Life was an element in the alife ecology. This ecology was also an ecology of bodies, the affects and pleasures that enabled the propagation of Life.

And what, we might ask, produced these effects? Whence came the pleasures of Life? Some, no doubt, were of a purely cognitive kind, but the emergence of complexity from the simple patterns on a grid is, for some, akin to the sublime. Christopher Langton, reporting on his close encounter with silicon, this time to journalist Steven Levy:

> The computer was running a long Life configuration, and Langton hadn't been monitoring it closely. Yet suddenly he felt a strong presence in the room. Something was there. He looked up, and the computer monitor showed an interesting configuration he hadn't previously encountered. "I crossed a threshold then," he recalls, "it was the first hint that there was a distinction between hardware and the behavior it would support... You had the feeling there was really something very deep here in this little artificial universe and its evolution through time."[27]

The depth that Langton encounters here, a depth associated if not identical to the vitality of the Life configuration, is beyond the screen. It is enabled by the simultaneous immersion in the world of Life and its disavowal, "Langton had not been monitoring it closely." I have written elsewhere of the way in which the autonomy and vitality of the alife creature is enabled by the structural blindness of this glance away: The power and vitality associated with the alife creature is directly related to its autonomy, and its autonomy occurs through the blockage of the machines, bodies, desires, and softwares that foster it. Hence, "life" is contained "in" this artificial universe, not in the (natural?) (uni?) universe. Just as identity is associated with an invisibility of the institutions and communities that enable it, so too does vitality seem to emerge only through the invisibility of its networks.

And yet, another response is also available here: Langton was looking at the alife body. That is, what Langton was looking at in the glance away was *not nothing*; it was the "external," material network of practices that enabled the uncanny and catastrophic movement on the screen: the emergence of a "distinction."

Just as the orchid images the wasp, so too do SimLife creatures image all too human desire, the desire for transcendental subjectivity of choice and purity of descendence. Such affective states are reproduced with each click of the mouse—Moofs help to reproduce the experience of transcendence as well as themselves. But the corporeal players of SimLife are also complicit in the "reproduction" of Simlife; the distribution of Simlife creatures, the deployment and articulation of their vitality, relies on the external wetware of human brains, their rhetorical softwares, their bodies and machines, just as the orchid is bound up with the wasp. Thus, Life, and by extension SimLife, is not a one-person game. Karl Sigmund claims that

> *Life* is not a two-person game like chess or checkers; neither is it a one-person game like patience or solitaire. It is a no-person game. One computer suffices. Even that is not strictly required, in fact, but it helps to follow the game. The role of human participants is reduced to that of onlookers. Apart from watching the game, one has just to decide from which position to start. All the rest proceeds by itself.[28]

Here I want to agree with Sigmund that Life, and all alife games, are "no-person" games. No person or subject reigns over SimLife creatures. They nonetheless do not proceed autonomously, by themselves. Alife works off of a distributed corporeality; the phenotype of the SimLife code, its necessary transduction apparatus as well as what it produces, includes the material, affective, and rhetorical ecologies within which they live. It relies on a Möbius body, one both within and without the screen, one at once inside the computer platform and outside it.

Thus as virtual witnesses to genomics, we form a rhizome with it. In this light, the transcendental discourse of "Topsite" that seems to cling to the space of the computer screen can be seen to be an effect of the screen among others, and not its cause. An instance of transspecies (indeed, transphylum) sexual selection, alife seduces with the allure of transcendence, so much silicon pheromone to traditional forms of vision and rationality. Yet in the midst of seduction, bodies become fluid and unpredictable: more than merely enacting the transcendental desire of the player, *SimLife* provokes a sense of complicity with a world beyond and around and of the screen, an exhilarating feeling of "depth," the sublimity of the chaotic multiplicity at play even in the simple *SimLife* world. The player becomes Moof, becomes trilobite, becomes plant, in the Simlife ecology, producing radically destratifying effects on the subjectivity of the user. Indeed, in some ways *SimLife* is a game of and about becoming, as the SimLife organism—including its interactive phenotype, the alife body of the user and his networks—only appears as recursive *blur.* The very materiality of the screen, the appearance of the SimLife organisms, seems to pulse. The creatures mutate, the user interacts, and nothing like a simple subjectivity reigns.[29]

But the plane of immanence, where moof, human, and silicon are all on the menu, abruptly crashes into transcendence, for genomics is not merely a technoscientific practice, it is a subjectivity machine. The constant choosing—as with Darwin's treatment of sexual selection, *SimLife* is also game of and about choice—reinscribes the difference humanity makes, collapsing and stalling the network into a node on the screen. The return from the glance away ends at the screen, a paradoxical "looking up" at Topsite, locating and rendering discrete the SimLife organism, transforming the user from a complicit element in an ecology to a choosing, viewing subject, an "onlooker," a witness of and not an ecstatic participant in the alife phenotype.

One final example helps make this virtual witnessing point. In the *SimLife* tutorial, the first action the user is directed to take in the SimLife world is to "draw your name in seeds." How does one "draw" a name? If this interaction operates as a vector of sexual selection, what desire does drawing a name in seeds solicit?

As a literal act of insemination, *SimLife*'s tutorial solicits the most obvious and allegedly biological vector of male desire: the desire to disperse ones "own" genetic information and therefore diversify, as the sociobiologists put it, a genetic profile.[30] And indeed, by "drawing" a name in seeds, one in fact writes oneself into a new phylum. With such a script, the user does indeed diversify their portfolio, expanding their reproduction franchise into the domain of silicon. No longer confined to the operation of human sexual reproduction, *SimLife* users now add a novel new ecology for their (foolish) hopes of immortality. Indeed, this ecology forms the substrate for that other wetware response to postvital living, uploading, a practice I will return to later in this text.

And yet one must ask the obvious and yet oft overlooked question: *How,* exactly, does one inseminate a machine? Only through a radical interaction and complicity—a submission to silicon—is the desire for insemination enacted. Allowing oneself to be solicited by a screen entails a seduction in which humans interact with, rather than act on, an entire bramble of technological infrastructure. Watching the seeds sprout, grow and die in something approaching the pattern of your name plays out the drama implicit in much of genomics. The drawing—an aesthetic rendering—of the signature, your signature, gives the effect of individual, human difference and control. You, after all, dominate that genie bottle on your desk, you populate it, design it. As the seeds sprout and plants grow and die on your simworld, "your" signature changes—bits fall out of a scrawled letter here and there, until becoming some other, barely legible, "word," then blending in with the messy background and frenzied movement of SimLife until yielding an unpredictable and transhuman growth. Hence the double gesture of transcendence and immanence: the user gets to inscribe and be overtaken by the genetic playground, SimLife. So too with the culture of genomics: the race to locate the genetic basis of everything simultaneous places control in the hands of "us" (doctors, biologists) even while it highlights our thoroughly masochistic submission to the inscriptions on our genes. That we might be infinitely complicit and entangled with our genes—neither masters nor victims, neither choosers nor chosen—evaporates under the spell of subjectivity, forestalling a more-than-genetic transformation, arresting the technological ensemble that could be a playground for becoming other than we are. Our challenge, as readers, writers, and players and not witnesses, is to hack this spell of transcendence with techniques of ecstasy.

T H R E E

Disciplined by the Future:

The Promising Bodies of Cryonics

Exactly what business are cryonicists in?

CryoNet online discussion group,
message 1734

IF THE contemporary hegemonic life sciences, with their incessant and productive substitution of DNA softwares for corporeal organisms, trouble our traditional conceptions of the organic body, *then* cryonics—the organized freezing of human bodies or heads in preparation for their revival—would seem to deploy a less elusive corporeality. A body is frozen, preserved, encapsulated against the entropic operations of time. Sealed off, stopped, suspended, nothing would seem so clearly isolated and autonomous as the cryonic body. Mute, it would seem outside the discursive networks that enable artificial life. Frozen, its very status would seem to forbid its connection to the intensities and affects which suffuse the artificial life phenotype and ribotype. In this sense, the cryonic corpse could be seen to be a memorial devoted to a nineteenth-century corporeality, a body of autonomy and will that has been put under a temporary, if uncertain, arrest.

Yet the cryonic corpse is a body of frenzy, an incessant smearing of the body over time. Neither alive nor dead, cryonic flesh organizes a massive discourse of maintenance and repair; even in its death, the body is becoming. Rather than simply ensuring the autonomy of a willing and paying subject through time, a great affirmation of the power of the Same, the cryonic body is saturated with difference. Reliant on a whole swarm of Others for its maintenance and its promise of revival, the cryonic body is not "technically" or accidentally immersed in a hospitality pool, along with its liquid nitrogen. Rather, in its very formulation and enactment, the promised body of cryonics is constitutively enthralled to the future.

In my investigation of the cryonic body that follows, I will attempt to highlight the transcendental production of an autonomous cryonic subject out of the tangled ecology that is the cryonic body, a discipline beyond living that ensures the constant identity of body and subject. At the same time, I hope to highlight the intensities produced in this economy of hospitality, intensities that fuel the futures market in bodies called cryonics.

Opening Up the Cryonic Body

An autopsy opens the body, displays the network of forces that produce the apparently unified organism, disturbing the claim that the body is one. The flow of blood, the elaborate network of organs, testify to the contingent production of life. Foucault, in *The Birth of the Clinic*, explicated the discursive possibility conditions for this "display." Not merely the scalpel cuts open the medical body; rhetorical formulations, silent and otherwise, made it possible for the opened body to "speak." If the body were to speak out of its silence, it had to be composed, patterned. *It had to be ordered.*

In that sense, then, an opening in or of the body relies fundamentally on performativity, an utterance or discourse that is itself an action, such as an order. That action, as J. L. Austin, Jacques Derrida, and Judith Butler have forcefully and differentially argued, cannot itself be rendered, articulated. It is a force that is itself to remain continually camouflaged as a condition of its production or emergence. In the catastrophe of an utterance, there suddenly emerges a command.

For cryonics, this is true from the beginning: "DO NOT AUTOPSY OR EMBALM!!" (CryoNet, message 1537).[1] This command—an action and a statement—has been inscribed on and in the refrigerator of a cryonics subscriber. It is not an incidental condition of the production of the cryonic body; it is an *initial* condition, the condition that must be true for cryonics to hold any promise. As a set of instructions that, its authors argue, must accompany the cryonic body, these commands and other recipes, rhetorical algorithms, have been characterized as a "vial of life" one ought to keep in the refrigerator.[2] The "vial of life" program was originally bundled with a set of medical instructions. Cryonics patients—and they must be patient—include cryonic directions within these medical recipes, these bits of exteriority that order a frozen, and not an opened, body.

Cryonic Softwares

Besides this stark example, though—the command that installs, even bootstraps, the cryonic body—there is a network of other rhetorical softwares that fuel the ongoing production of a cryonic corpse, a set of rhetorical, disciplinary operations that Deleuze and Guattari have characterized as "order words."[3] These rhetorics take the form of a promising machine or organization. Cryonics *is* a promise, a promise to revive that is itself continually at risk, in exposure. Something other than the present, something yet to come, insures the cryonic body.[4] Cryonics cannot stand to be alone; it is itself networked with something other, even if that other be nothing but futurity. Productive, in frenzied motion, the frozen body does not stand still. It moves along a constant reinscription of the promise, a constant promise to promise. It is in part the promise of the machine:

> If someone was to place a regular amount each month into savings, and when sufficient funds were available buy some conservative high technology stocks, or maybe a mutual fund specializing in technology, then if growth does not occur this is because technology has failed in some way, and this same failure will defeat cryonics as well. Therefore the system is self-fulfilling: No technology growth = no cryonics, and you have no funds. High technology growth = rising funds and successful cryonics, which you can afford. (CryoNet, message 4856)

This statement—sampled from an online cryonic discussion group, CryoNet—highlights the way in which the promise of cryonics is intimate with, if not identical to, the promise of a certain set of technologies, mostly nanotechnology. But it also indexes the way in which cryonics is dependent upon a marketing of the future, or at least a futures market. If high technology stocks are, at least in part, promises to deliver a set of knowledges and techniques in the future, then cryonics is a futures market in bodies.[5] "One living body to be delivered at some incalculable date." But it is not merely any living body to be delivered or named later. It is a living body coexistent with the subjectivity of the cryonics subscriber; typically this return of the subject, the fulfillment of the promise of cryonics, is thematized as the patient who "awakens." In *Cryonics: Reaching for Tomorrow*, a publication put out by the Alcor corporation, one of the largest cryonics organizations,

the speculative scenario for recovery is a science fiction story emphasizing the identity of the client. After a technical description of the (nanotechnological) means of
revival that is voiced in the third person, the patient reports as the one who says "I"
repeatedly:

> I awoke to a gentle touch on my arm. I opened my eyes and saw my friend
> Ken Phelan. He was wearing a white lab coat. I looked around me and saw
> that I was in a hospital room.[6]

Unlike Kafka's Joseph K. of *The Trial*,[7] who awakens to find that he has become
something other, a criminal whose crimes are unknown to him, the cryonics patient
is promised a self that will persist even through the sudden avalanche of identity
called "awakening." I am still I. Friends and family have become healthier, wealthier,
but not different. Subjectivity persists in death in a manner impossible in life; if identity is a set of becomings, it is only in becoming-frozen that becoming itself is frozen.

The Other Side of the Freezer

Strangely, this promise of cryonics is also the promise of hospitality, a promise which
for philosopher Emmanuel Levinas is fundamentally ethical, as it involves the
subject with an inarticulate other.[8] Posting the simple but impossible question—
What should be done?—the Alcor corporation, one of the largest extant organizations
that practice cryonics, thematizes this fundamental alterity as the "relative problem":

> With twenty-seven patients in suspension, Alcor has had quite a lot of expe-
> rience dealing with relatives of suspension patients. Even with friendly, co-
> operative relatives there can be many problems. One of the biggest problems
> that most people don't consider is that even friendly relatives (on average)
> don't know much about cryonics. They don't understand what the procedures
> include, what kinds of cooperation we need from hospital, nursing home, or
> family, and they don't understand why we hang around the patient for several
> days, asking nosy questions about the patient's blood pressure, temperature,
> enzyme levels, etc. We need this information to estimate when the patient
> might deanimate, so we 1) don't waste time and money by getting to the pa-
> tient's side days early, and 2) so we can be prepared and ready to act right
> away when the physician (or nurse, in some cases) pronounces death. This
> prediction can be remarkably difficult in the best of circumstances. (CryoNet,
> message 2203.1)

Thus even as cryonicists demand to "take control of their lives" and to "rage against
death," they are in a position of exposure; contingency could intervene at any

moment. The technical preparation is constantly implicated in the understanding of an other, others for whom the act of gathering "information" becomes an act of intervention.

"On this side of the freezer," what cryonics pioneer Robert Ettinger calls his life so far,[9] the problem of the other is primarily figured as a problem of information. As such, the "relative problem" foregrounds cryonics' reliance on a notion of human interaction that is founded primarily on an information theoretic paradigm of communication: Any resistance to cryonics, and any inhospitable response to the hospitality machine of cryonics, is regarded as a lack of knowledge, not an excess of difference. We will return to this notion of cryonics' debt to the rhetorics and concepts of information, but for now I just want to mark the inability of cryonics to think the perhaps impossible-to-codify difference that people other than cryonicists are. This inability is not merely a side effect of cryonics; it enables it, in so far as the very project of archiving bodies and subjects relies upon the possibility of stabilizing information over time. And, as we shall see, it is precisely difference and its inarticulability that threatens the stability of information.

The "relative problem" also highlights the fact that the cryonic body is not simply a dead body: any cryonic subjectivity is always already becoming-frozen. She has entered into a relation with a future, a cryonics organization, and a massive realm of incalculable contingency that must be continually managed, even disciplined. Hence the amulet worn around the neck of the cryonic subscriber is a node in a massive network of practices: helicopters for body transport, the promised goodwill of relatives, the future itself, all spiral out from a necklace.

These camouflaged—not hidden—background conditions or contingencies compose what Deleuze and Guattari, after Henri Bergson, render as the "virtual." Thus one response to the search engine query this segment begins with—"*What business are cryonicists in?*"—is: the virtual body. This is not simply the future body and subject that will reappear at the vanishing point of the future; it is the set of rhetorical softwares that enable the production and fantasmatic actualization of a futural subject at the site of the body's smooth space of inarticulation, its future of *static.* Deleuze and Guattari note that these "incorporeal" articulations are nonetheless networked with corporeality:

> Incorporeal attributes apply to bodies and only to bodies. . . . Their purpose is not to describe or represent bodies: bodies already have proper qualities, actions and passions, souls, in short forms which are themselves bodies. Representations are bodies too! . . . We cannot even say that the body or state of

things is the "referent" of the sign. In expressing the noncorporeal attribute, and by that token attributing it to the body, one is not representing or referring but *intervening* in a way; it is a speech act.[10]

It is precisely this notion of "intervening" that I want to highlight here; the cryonic body is not autonomous. Alive or dead, it is *on the line*, connected to the rhetorical, material practices of "information" collection and dissemination that make up the cryonics community. While subjectivity may be enabled only via an alterity, which itself is inarticulable, this is particularly obvious with cryonic subjectivity. The production of a cryonic subject—alive, dead, or in the uncanny space of suspension—is possible only on the basis of a Möbius body, a body both within and without the capsule of liquid nitrogen, inside and outside of time. Here the cryonic body exemplifies Levinas's observations about subjectivity: "Subjectivity realizes these impossible exigencies—the astonishing feat of containing more than it is possible to contain."[11] The cryonic subject, alive or dead, thus "contains" more than itself; as a body with an ongoing subjectivity, the cryonic body is oddly shaped, as it contains its future. It depends on the boundless need for an ongoing promise, a promise to preserve the body, name, and project of the cryonic subject.

Thus, for example, the cryonic subject can actually own in the future, containing assets in more than the present, as in this example taken from a Usenet Frequently Asked Question list (FAQ) for cryonics.

> How can I pay for my own revival and rehabilitation, and keep some of my financial assets after revival?

> The Reanimation Foundation is set up to enable you to "take it with you" and provide financial support for your reanimation, reeducation, and reentry. It is based in Liechtenstein, which does not have a Rule Against Perpetuities, and thus allows financial assets to be owned by a person long after the person is declared legally dead.[12]

What this transaction alludes to is the instantiation of a financial subject on the other side of the freezer. Installed in the present, this is an annuity of anticipation, a prolepsis of capital. But in the logic of capital—and cryonics is a body of capital, a chunk of capital's body—assets, even futural ones, must be offset by obligations, an obligation not just to the other, but an obligation to capital, a logic of capital that contains more than the present. In the futures market of bodies called cryonics, the immortality of capital itself is at stake in a futures market of markets. More than the body must survive its long hiatus, more than the goodwill and solvency of the cryonics corporation is at stake; capital itself must persist.

The cryonic subject also owns the secret of her subjectivity; according to the FAQ, she owns her own secrecy:

> Is Walt Disney frozen?
>
> No. There was a time when all of the cryonics organizations would tell you this. Since then Alcor (possibly among others) has realized that if they admit when an individual is not frozen, then it is possible to infer by elimination who is frozen, which they have in many cases agreed to keep secret. Thus Alcor will no longer say anything informative about whether Disney was frozen. Nevertheless, Disney is not frozen.[13]

Even in death, fame does its work, opening graves with rumor and gossip. One can only guess at the status of this response, mired as it is in a rhetorical recursion without end. The contradictory nature of this response—we won't say, we will say, in the saying we are not saying, in the not saying we are saying—rhetorically mimes the "suspension" of cryonics. The status of the cryonic subject is indefinitely deferred, as Alcor preserves the possibility of the cryonic subject precisely through a production of rhetorical undecidability.[14]

This undecidability, useful though it is in fostering the cryonic subject's ownership of her secrets, haunts any ownership of the body. Alcor president Steven Bridge details the difficulty of "owning" a body of the future here:

> Decide who owns your body. If your spouse owns your body, then you'll have to settle for no cryonics. If you own your body, then do cryonics and try to make your spouse deal with the resulting negative emotions instead of you. If you and your spouse have shared ownership of your body, then the two of you will have interminable arguments about it; I therefore disrecommend this option. "you own your own body" is legally impossible after you are declared legally dead; you can't own anything. Actually, the laws of the various states determine who owns your body—or at least who has the authority to dispose of it. For you to *change* that authority, you have to do something active. This appears to be solved by the Anatomical Donation to Alcor. *Persuading* your spouse that this solves the problem is an entirely different story, of course. A hostile relative or spouse can effectively prevent your suspension or delay it for a very long time by simply not informing Alcor that you are dead or dying. Of course, if they don't even know you are signed up,

they won't call Alcor either. We don't have a radio-powered heart monitor attached to those Alcor tags, folks. (CryoNet, message 2203.1)

What's remarkable about this response—among other things—is that Bridge high-lights not simply the difficulty of owning a future body, but the problematic nature of the speech act associated with it. "Interminable arguments"—such as those de-ployed by Alcor in the production of Disney's "secret"—threaten the algorithmic production of a cryonic body and subject. The body—coded as a "donation" of organs to Alcor—nonetheless resists any complete coding, any final rhetorical and linguistic management of contingency. "Persuading" the other entails something other than the transfer of information or its legal inscription: it is a mobilization of rhetorical force that itself is beyond, before, or in excess of the code; it is "an entirely different story," a story of difference.

This massive debt to alterity would seem to threaten the cryonic project of personal, subjective immortality. For the promise of revival, what cryonicist Robert Ettinger has dubbed the "prospect of immortality," does not resound with the immortality of the community; it sounds the note of identity: "me me me me." It does not concern or promise the arrest of time, the erasure of the vanishing point, for the community—time will march on, difference will happen. It's just that every-body will live forever, everybody with the commitment to cryonics. This is a crucial element in the narratives that thematize cryonics as time travel: history "moves on" with astonishing speed, in contrast to the mute indifference of cryonic subjects.

Becoming Code

How is this threat—the tension between the absolute difference of historical, tech-nological change and the identity of the cryonic subject that is enabled with it—managed? A return to Bridge's discussion of the "ownership" of the body provides us with one answer: "Decide who owns your body." This decision—its possibility bound up with the rhetorical formulation that one "has" a body rather than "is" one—indexes the way in which the commitment to cryonics is also a commitment to the notion of the body as code. For the "decision" to own one's body, whatever the rhetorical and agonistic trajectory that leads up to it, relies on its inscription into a code, a legal code. This speech act—the utterance marks the body with ownership, intervenes in its future—depends upon a disciplinary network of organ donation and "maintenance" that ultimately relies on the possibility of transmitting coded commands over time, *in a code outside of time*. This coding of the body over time, the program of cryonic production and revival, is continually threatened by the insufficiency of the coded body: "we don't have a radio-powered heart monitor attached to those Alcor tags, folks." This perceived insufficiency of the coded body produces a constant, nearly frenzied attempt to code the cryonic subject, a coding that is in part carried out by the online cryonics community.

A particularly persistent and sometimes acrimonious exchange on CryoNet concerns the status of a personal "archive," a bundle of information that accompanies the individual cryonic body in order to stoke the memories of the revived patient. Here the physicality and limits of coding come to the fore, as the coding of the cryonic subject—in anticipation of a future reading—becomes a near constant concern in the present. The very anticipations that concern the production and revival of a cryonic *body*—the need to produce a code outside of time, or at least one readable in the future—haunt the archive as well:

> The issue is whether you want to put out data and hope that someone can reconstruct it, or whether you want to do the best job of storage you can do. This is much like when people doing cryonics don't simply flash freeze people and hope that the technology will be good enough someday—they try to do the best possible preservation so as to minimize the work people have to do in the future and maximize odds that the person can be fixed. Well, are you trying to maximize the odds that your data will go through, or not? Something that survives to the future without the need for reconstruction has far better odds of being read than something that requires reconstruction. (Cryo-Net, message 5663)

Just how these "odds" can be evaluated, of course, remains to be seen, but for most cryonicists on the net, the more discipline that can be exerted on the data—the more it can be insulated, if not encapsulated, from the operations of time—the better. Indeed, the constant anticipation of a need for memories in the future seems to be the very "purpose" of the present, as in this rather Borgesian scenario:

> If I had huge amounts of money I might well hire a team to follow me about videoing my life and taking notes about every little thing I did and said. If I had more stamina I might write a very detailed diary myself and have it stored, which would be a relatively cheap option, and actually would be pretty useful, given the way my life has gone so far. A diary might even bring in some revenue this side of the dewar. (CryoNet, message 5620)[15]

The very volume and intensity of the exchanges concerning the archiving of personal information indexes the possibility that as "informational beings," cryonicists are almost as concerned with the survival of "external" information as "internal" information. Indeed, in the cases cited above, the very difference between the two seems to disappear, as does the distinction between "life" and "preparation for revival." Still, Mike Darwin, a well-known and controversial cryonicist with a penchant for pragmatics and polemic, sees the difference between archive and flesh as significant indeed. Addressing a discussion of compact disks as appropriate archival media, Darwin writes,

> You guys are worried about decoding CDs. Talk about dumb and dumber. The information on how to do *that* exists; algorithms are used in existing players; *get them*. The principles are *simple*. What gives me a belly laugh is that while you all agonize over CD formats being accessible, adhesives decaying, metal pitting, etc. you seem quite happy to have your brains all hashed up and then blithely expect someone to decode *that* mess, *and* without the instruction manual. Talk about misplaced priorities. (CryoNet, message 5576)

But perhaps Darwin's intervention, insofar as it relies on the alleged stupidity of the discussion, misses a crucial point: for many cryonicists, apprehension about the stability of archival media over time *is* in some sense a discussion about the possibility of revival in the future; the very operations that disturb the possibility of an archive—decay, the sheer difference of the future ("will they have CD players?!"), fire ("At least while you are animate…a personal archive refrigerator is helpful in many ways." [CryoNet, message 5591]) also threaten cryonic bodies. Indeed, reflection on the common project of archiving texts, photos, and bodies leads to the remarkable news that CryoNet should itself be part of the "gateway to the future."

> One way to save some material about yourself is to publish. Published material is widely disseminated. There is a good chance that at least one copy will survive. Which brings us to CryoNet Digest. What is published on CryoNet Digest is certainly stored in a many places (at least for a while), but is there a plan for systematically archiving the pearls of information, wit, and wisdom we see everyday on the net? (CryoNet, message 5564)

Thus the "information" deployed in cryonics is anything but immaterial. It is a constant commitment to the future in the present, a commitment with material consequences that extend well beyond the expense and oddity of compiling one's own archive for the future. Consider the surveillance system suggested by Mike Darwin in response to the fear of an unreported death. After the death of a cryonics subscriber who was not found until he had decomposed for three days (he was writing his thesis in philosophy), rendering cryonic preservation impossible, Darwin suggested the following procedure:

> Standard alarm components that sense motion can be mounted throughout a house. When you come into the house you turn the keypad on. If there is no motion for say more than 30 minutes, the systems sounds a local alarm which can be manually overridden in say 5 minutes. If it is *not* overriden it can dial 911 or the cryonics group or a neighbor, etc. And, anticipating a likely objection: sleeping people move; they breathe, they toss they turn, etc. A more sensitive detector is needed for the bed, but they exist and are inexpensive; kinds for mattresses and kinds for overhead mounting (infrared). By the bed is a cancel switch and a panic button. Panic button triggers that can be worn on necklaces are commercially available that can be tied into the system for several hundred dollars more ... When you leave, you key out. True it won't help if you die away from home, but so far exactly *none* of the cases of unattended SCD have occurred anywhere else. And this is a *good* start. (CryoNet, message 5576)

Thus cryonics, more than a treatment of a legally dead body, becomes a disciplinary operation of the living. The opposition between "life" and "death" is undone not only on the basis of a revival yet to come, but in a present that becomes a constant rehearsal for the future, a future not quite here, not quite dead.

This remarkable commitment or discipline extends to the rhetorical possibility conditions that have made the cryonic body thinkable. Of course, cryonics is itself a heterogeneous field, with multiple sites of contestation and difference. Indeed, the CryoNet electronic mailing list and Usenet group that form the archive for my research are composed of these differences. One of the most active sites of difference in the cryonics community is the "uploading" argument popularized in part by computer scientist Hans Moravec.

> It is easy to imagine human thought freed from bondage to a mortal body ... A computation in progress—what we can reasonably call a computer's thought process—can be halted in midstep and transferred, as a program and data read out of the machine's memory, into a physically different computer, there to resume as though nothing had happened. Imagine that a human mind might be freed from its brain in some analogous (if much more technically challenging) way.[16]

By contrast, cryonicists such as Robert Ettinger argue for the specificity of corporeal embodiment. But both the defenders of the "body" and the aficionados of silicon share a preunderstanding: they articulate the body as pure code. As Microsoft founder and media event Bill Gates told biographers Stephen Manes and Paul Andrews, both the body and consciousness become a "software problem."[17]

It is clear enough that the implosions I discussed above and elsewhere, the conflation of life and information, are crucial rhetorical vectors for the uploaders; "copying" humans onto a floppy disk is in some sense merely a logical conclusion of biologist Richard Dawkins's claim that DNA is a floppy disk. Perhaps it is less apparent, however, that the promise accorded corporeal revival is itself tied to the ability of the body—and not merely its subjectivity as owner, owned—to become code.

Becoming Crypt

Ralph Merkle, a nanotechnology researcher at Xerox PARC (Palo Alto Research Center) and a prominent cryonicist, underscores the coded nature of the cryonic body in his research essay, "Cryonics, Cryptography, and Maximum Likelihood Estimation."[18] While I must again highlight the heterogeneous nature of the cryonics community, it is also true that Merkle occupies an important place in the world of cryonics theory. "Cryonics, Cryptography, and Maximum Likelihood Estimation" thus offers us one of the most talked-about and respected "revival scenarios," the part-technical, part-fictional texts—literally, science fiction—that are the currency of cryonics' claim to a futures market in bodies.

The pun, of course, is obvious: "Crypt" refers not just to the notion of a cipher or a secret; it also names the very object that cryonics aims to open up, to empty. It is upon this pun, I will try to show, that Merkle's revival scenario hinges. After a long discussion of the difficulties of cryonic revival, Merkle recasts the question, and, indeed, the body:

> So the question of whether or not we can revive a person who has been frozen can be transformed into a new question: can we cryptanalyze the "encrypted message" that is the frozen person and deduce the "plain text" which is the healthy person that we wish to restore? Are the "cryptographic transformations" applied during freezing sufficient to thwart our cryptanalytic skill for all time?[19]

What makes possible this transformation of the question of revival? While there are a number and a network of presuppositions that make the translation of the body into code plausible, the characterization of the body into nothing but an extension of a code, DNA, marks out one of the more obvious possibilities. For surely, if the body is itself nothing but a lumbering robot, a machine dedicated to the continuation and propagation of selfish DNA, then it can also itself be rendered *as* code. If molecular biology fostered a forgetting of the body and a remembering of code, DNA, then cryonics enables a return of the body, the body as code.

But the analogy between the cryonic corpse and a cipher remains suspended in difference. For while a cipher represents a systematic transformation of a plain text, a text known by some subject in the past, the cryonic body signifies an absence of such knowledge. A revival of the cryonic body relies on what, precisely, those of us in the past do not know, and as such cannot inscribe.

For example: Any coding of the subjectivity of a cryonic subject would demand a knowledge of subjectivity in the present, a precise characterization of the answer to the question, Who is the cryonic subject? Any such answer would need, at the very least, to meet the challenge of the Turing Test: when faced with their own subjectivity, the cryonic subject would not be able to tell who was talking to whom, whether or not she was simply talking with herself. Needless to say, such a precise coding of subjectivity remains a promise, as the subject is precisely the one who doesn't know who she is talking to; transcendence is impossible at that location. One could never know if one had succeeded or not; any answer would be haunted by the fact that one might be speaking with an other, if not the other, that one might simply be speaking to oneself.[20]

There may indeed be more material limitations on this coding than technological ones. For if subjectivity—as Butler, for example, insists—is a process of iteration and not an essential attribute, then the coding of a subjectivity becomes not just implausible, but impossible. For such subjectivity forbids the possibility of a transcendental position on its "own" subjectivity; such a position would be merely one more iteration of the subject, not one that arrests the becoming of identity into being once and for all.

Of course, something other than a simple subjectivity may operate here: subjectivity's double. If the iterative process of subjectivity shatters any possibility of a simply mimetic or disclosive articulation of the crypt/cipher of subjectivity, then the production of a cryonic subject in Merkle's scenario summons the uncanny fact that "subjects," as iterative becomings, aren't even identical to themselves.[21] This uncanniness, while troubling the production of immortality as an endless rendering of the same—"me me me"—fosters an encounter with becomings other than freezing, becoming *static*. Indeed, this "static" produced by becoming-frozen tunes our ears to the multiplicity at play in the disciplines and machines of identity; Merkle's deployment of the machine analogy reminds us that contemporary subjectivities are precisely the effect of a contingent assemblage of machines, an inhuman ecology that lives off difference. Perhaps in this regard, Walt Disney's body and subjectivity is more illustrative of postmodern subjectivity than we would like to admit: alive, dead, or animated, subjectivity demands difference. More on difference and its demands later, I promise.

It may seem that the body represents a more straightforward cryptographic possibility. For if subjectivity resists a total inscription—if it's as much an element in the discourse suffusing the subject as an attribute of the subject "herself"—then certainly the body is its own best record. But the body is already a ruined text, unreadable even in the present. The cryptographic model of revival, the transformation of the body into code, would seem to demand an "original" body, a plain text that would order, irreducibly, the entropic transformations of time. The remains of a body, then, would harbor information, information that testified more to a future life than the cause of death. Merkle compares this persistence of the corporeal media to the national security challenges posed by the difficulties of erasure:

> As experience with erasing top secret media has demonstrated, it's hard to get rid of information when sophisticated means of data recovery are employed. And we'll have very sophisticated means of data recovery available to us in the future.[22]

But again, this notion that a secret can be retrieved even after its destruction depends upon an original existence of the secret. A secret is a knowledge that is hidden, and the scale and depth of knowledge of an individual body that Merkle's revival scenario demands is not hidden—it is simply not known. Perhaps it is not known for good reason: like identity, the body entails not so much being as an unpredictably complex becoming, a dynamic, differentiating event and not a static object. Thus the revival scenario depends not so much on knowledge in the future as in the present—a knowledge of what bodies "are" and what subjectivity "is." By analogy, our reading of the title of Merkle's text—is it a pun or isn't it?—remains undecidable in the present, as it entails a dynamic event of rhetorical resonance and not a simple inscription of information. No amount of cryptographic skill, now or in the future, will tell us what "crypt" "really" "means."

Still, while the analogy of cryptography may not hold insofar as one compares the body to a scrambled plain text, we may still preserve the possibilities that Merkle's cryonic revival scheme engenders. For the production of a cryonic body does not depend so much on the semantic fitness of its rhetorics as on the pragmatic formations of power and knowledge that they make possible. Molecular biology was made plausible and enacted on the basis of equally problematic tropes; the genetic "code," the notion of genetic "information," among others, were rhetorical softwares that were constitutive of molecular biology even as they opened up lacunae in its "foundations." It is possible that the cryptographic body, networked with the material, rhetorical practices that suffuse it, will make plausible equally unprecedented knowledges. The work of thinking is to be something other than witnesses to this cryptographic body. Indeed, perhaps we must render the cryonic body and its attendant technological transformations in terms of a tool box for becoming other, recipes that will cultivate a future of some difference.

Speeding to a Freezing, Anticipatory Speeding

If I have highlighted the rhetorical dependencies and difficulties that enable and trouble the production of cryonic subjects and bodies, I do so not to critique the project of cryonics. The attempt to preserve bodies and subjects provides us with a map of contemporary biotechnology—what bodies and subjects can be produced, today?—and it is in this spirit that I research and study the rhetoric of cryonics. But cryonics, of course, is also concerned with what bodies and subjects might become. For if the production of cryonic subjects and bodies seems to depend on a vanishing point of total knowledge in the future, it also produces a set of effects on the becoming that is the body. What effect does cryonics have in the present? To return to the question with which we began, what business is cryonics in, now?

If cryonics is in the business of producing a virtual body, it is a body of anticipation. For if the corpse has been, in our memory, a body of memory, cryonics transforms the corpse into an anticipated body. The frenzy that now greets the death of a cryonicist is a measure of this transformation. Robert F. Nelson describes this frenzied anticipation in *We Froze the First Man: The Startling True Story of the First Great Step toward Human Immortality*. Nelson, learning of the death of the first cryonics "patient," speeds toward the death scene:

> If there is a law I didn't break driving to Glendale that afternoon, it hasn't been written yet. I put my foot to the floor and proceeded at breakneck speed, oblivious to red lights, stop signs, pedestrians, only vaguely aware of the squealing of brakes all around me.[23]

The endless process of disciplining, preparing the cryonic body, accelerates. By contrast to the sense of resignation, grief, and closure that attends most contemporary death scenes—"We did all we could"—clinical death marks only the acceleration of an endless, proleptic labor. If the body is to become code, to become readable by and in the future, then the cryonics organization must continually anticipate that reading. The moment of death marks the acceleration, and not the arrest, of the treatment of the cryonic body, becoming frozen, becoming code. As such, the trip to Glendale marks a new vision of a favorite sitcom tableau: the dangerous and frenzied journey demanded by a laboring body, the strange complicity of the speeding car in the birth of a human.

The anticipation above is not simply a physical rush toward the dead body, it is also the anticipation of a law that will not have been broken, a future law that nonetheless holds sway over the present. "If there is a law I didn't break driving to Glendale that afternoon, it hasn't been written yet." Nelson's initial swerve toward cryonics breaks the law(s) of the present, but this present, the cryonic present, is understandable only from the standpoint of the future, a law of the future that Nelson obeys rather than transgresses. This law—unwritten, yet perhaps writable in, with, the future—is perhaps the law that emerges with the cryonic body, the law of prolepsis that cryonicists anticipate. They anticipate a future of anticipation, a future for whom the past *is* nothing other than the anticipation of the present. The cryonic subject only wants what science already has for itself—*a law that will hold for all time:* that life, and not just matter, should be neither created nor destroyed.[24]

As machines for the production of identity in the future, the cryonic assemblage is a reef of signification and subjectivity, a franchise on the plane of transcendence. As such, cryonics sends the future the worst symptom of the present, a frozen pocket of interiority. And yet cryonic desire is also something other than the desire to be a subject; it is an *experiment* in becoming a scientific object. Rather than merely observing the infinite operations of scientific laws on that stable background, the "universe," cryonicists desire to become laws themselves, endless repetitions that hold for all time. Cryonics fosters a body continually anticipating a judgment of the future, and in its anticipation it eludes containment by any univocal judgment, as the status of the cryonic body is anything but decidable; it is to come.

The cryonic body is not, then, simply the instantiation of bad conscience taken to its limit, a mobilization of power that allows flesh to leap out of the hell of time, to make time stop, to finally constitute the body as a pause function in a nonstop negentropy machine. It is a body composed of speed that experiments *on* and *with* time, a production not of stasis but of *waiting*. Cryonics is not so much an attempt to avoid the great lack of death as the purchase of a certain style of death, a death of some difference. Something other than mourning clings to the cryonic body, an unthinkable and incalculable future that, of course, in its formulation is nothing other than the articulation of the present. Hence the oddity of the cryonic body: As with the rhetorical suspension of Disney's subjectivity—Is he or isn't he?—the cryonic body is a production of irreducible difference, flickering in and out of time, in and out of possibility with each new articulation of its description and treatment. Perhaps this explains the massive discourse produced by cryonics: like sexuality for the Victorians, the cryonic body must be incessantly and continually described, judged, evaluated, explained. This intensive evaluation—as in the endless surveillance discussed by Darwin above—provokes a rhetorical crisis: How to evaluate a transaction that takes place in the future?

> What is the value of a life-saving surgical operation? What is the value of being cryonically suspended? It isn't infinite—there isn't yet enough money in the world to give every ill person every possible medical treatment that might help. But wherever the line is drawn, some people will be condemned to death. Should government be the one to draw this line? . . . Or should it be triage via waiting list, in which if 100 people need a life-saving treatment, and the government has allocated resources to treat 60, the 100 will be forced to wait until 40 of them have died? Or perhaps a lottery? Or a TV show whose viewers vote by phone on which patient's pleading was the most pathetic and moving? (CryoNet, message 5075)

It is the very impossibility of evaluating such a future event that constitutes the cryonic body. As a production of anticipation, cryonics is oriented around an invisible, unverifiable point of comparison, a space and time of impossible evaluation, a moment of sheer difference, the difference of revival. The constitutive ambiguity of cryonics, rather than a lack that threatens the coherence of the decision—"How could I make a decision based on an absence of knowledge?"—produces an economy of excess, an economy in which there can never be enough evaluation.

> Perhaps most cryobiologists perceive cryonics advocates as implicitly claiming a probability of revival of 50–90%, and therefore see us as frauds, a perception they wouldn't have if they thought we were only claiming a 1% chance of success. If only one or two cryonics folks have published specific probability predictions, perhaps we should do more public probability estimation before complaining that the other side doesn't do any. My home page announces I think I have at least a 5% chance of revival. Have you big-name cryonics folks announced your estimated probability of revival? (CryoNet, message 5627)

While the "value" of a cryonics contract is not infinite, the task of *evaluating* it is. We could characterize this endless process of evaluating cryonics as a failure, a *differend* whose resolution relies on premises and technologies from the future. But rather than thematizing this as a loss—which cryonics, faced with death, can't do—I would like to briefly highlight the effects that this infinite evaluation produces in the present. The evaluation of a cryonics contract—lining up capital with bodies in a coherent fashion—also produces what Brian Rotman has characterized as "being-Besides-oneself," a doubling of subjectivity he associates with the emergence of parallel processes, a parallel "subjectivity."[25] For the evaluation of cryonics is, to perversely follow Shapin and Schaffer, a kind of virtual witnessing of one's own death. More than a desire to live forever or to avoid death, cryonics fosters the desire to remain present at one's own death, to double one's self, to go parallel.

Thus the frenzy that inheres in the cryonic body that surrounds it like liquid nitrogen is exuded by the double tension of cryonics, the tension of the double. Oozing with transcendence—the desire to master one's own death, to see it—cryonic bodies also secrete immanence, a decoupling of bodies from loss, the production of something other than a simply doubled subjectivity: a parallel subject that shatters the unity of the "one" who is becoming-frozen. Destratifying the subject by launching it through and out of time, cryonics also enables a scientific body, a body that would itself, like the truth, be untarnished by finitude. The coming of the cryonic body—and it *is* coming, more about that later—resonates with the doubleness of the plane of consistency and transcendence. This doubleness is the space of the future, and perhaps contemporary, body, a body within and without time, within and without the computer, a Möbius body.

Deriving the Cryonic Body

In mapping out the possibilities of the cryonic body, I have tended to focus on the ways in which becoming-frozen is thinkable, articulable in the material, rhetorical ecology of cryonics. But beyond the networks of rhetorical and material practices that are complicit with the emergence of the cryonic body, it would be some measure of contemporary distributions of the body and desire to determine the *naturalization* of cryonics, the investment of cryonics with a telos toward which we have been striving all along, as a kind of end point of the anthropic principle, an algorithm that suggests that the universe exists through and for the purposive gaze of human subjects, subjects whose very purpose is to be frozen, to pause, to slow down. Indeed, I am often struck by the ease with which many people, myself included, come to regard cryonics as a natural, and almost inevitable, desire. In its slow discursive bleed from science fiction to science/fiction, cryonics seems to have left the very fantasmatic luster that suspends, like so many masochists, belief.

The shedding of this fantastic resonance is complicit with a new figuration of the future itself, a writing practice that is quite literally beyond value, beyond the metaphysics of valuation, the derivative. Derivatives, as writing practices that in some sense write *on* or *with* future, dramatize the articulation of the cryonics contract as a contract made out of something yet to come. As such, the rhetorical softwares that traverse cryonics and certain flavors of derivatives also map out the emergence of something other than capital, capital's *advance*. But what, to ask a question in advance, is a derivative?

Brian Rotman describes the transformations that overtook mathematics, painting, and finance with the introduction of zero as a "metasign," a sign that, semiotically speaking, stands for the *absence* of other signs. Zero, Rotman writes, "is thus a sign about signs, a meta-sign, whose meaning as a name lies in the way it indicates the absence of the names 1, 2, ..., 9."[26] Beyond his interrogation of the semiotic operations of mathematics, Rotman locates isomorphic discursive transformations in painting and finance. Rotman demonstrates that the vanishing point in painting, as correlative of zero in a visual grammar, organizes the possibility of a viewing subject, a subject for whom the perspectival painting becomes a talisman that provokes the artist's point of view. More than a simple inscription of a prior subjectivity into the medium of paint, light, and canvas, then, the vanishing point is a discursive machine that *orders* or installs a subjectivity through the recursive operation of the meta-sign—a sign of signs that provokes something like reflexivity, as perspective offers the spectator the possibility of objectifying himself, the means of perceiving himself, from the outside, as a unitary seeing subject, since each image makes a deictic declaration: this is how I see (or would see) some real or imagined scene from this particular spot at this particular moment in time.[27]

This achievement of the vanishing point, a rhetorical articulation of nowhere that, like zero, signifies nothing, makes possible the evaluation of the painting; as the possibility of objectification, the vanishing point makes it possible for the viewer to understand how *others* would see the painting as well, a recursive operation that enables exchange, the exchange of evaluation.

Given this description of the semiotic operations of the vanishing point as productive of a certain style of viewer in the aesthetic encounter or transaction, it is easy to see the rhetorical ecologies enabled by the shifting status of

monetary signs. For example, the substitution of the sign of currency as a representational machine for transmitting the value of gold increases the velocity of any given
transaction—no species metal must be delivered for the rhetorical moment of a
transaction to occur. The act of evaluation could now occur at a distance—in space
or time—from its transcendental guarantee, gold. "Reference" or "the representation
effect" becomes one of the necessary components of any economic transaction under
such a regime, and, as Rotman points out, any such representational regime is haunted
by the possibility of forgery. Thus the demand for not simply species metal that currency "refers" to, but of a subject—such as a state or a corporation—that is the absent "evaluator" of any given currency. To paraphrase Rotman, "this is how I, from
the outside of this transaction, see this currency." If the degree zero of the vanishing
point enabled the spatial depth of painting, the virtual enactment of the viewing
subject, the rhetorical economy of currency promised value at a distance, both in
time and space.

When this regime of reference was finally evaluated and aban-
doned at Bretton Woods in 1972, the ghost of the "evaluator" in this semiotic economy was gone. No longer was any "subject" outside the regime of transaction evaluating currency, giving it its perspective, the eye of god that adorns the back of U.S.
currency. Instead, the evaluation of U.S. currency—and, by extension, the world
financial markets—became a practice embedded in the future. The "outside observer" that guaranteed any given transaction was immanent to the market, as the
eye of god moved into the future, a futures market. Futures markets took over the
role of absent observer implied in every transaction, as currency markets collectively
declared, "This is how I, outside of your transaction, evaluate money." It is as if the
financial vanishing point itself vanished, invisibly perched on the other side of the
present, guaranteeing currency from the future.

As you have probably anticipated, derivatives continue and accel-
erate this shift of the financial vanishing point. While the term "derivative" is itself
imprecise—"the term derivatives is not really well defined. It has become a catch-all
generic term that has been used to include all types of new (and some old) financial
instruments"[28]—a small percentage of the futures markets is comprised by futures
contracts that cannot themselves be resold, or "marked to market." Without the
market as the absent but available "evaluator" in this rhetorical economy, so-called
exotic derivatives mark yet another shift in the financial vanishing point, the collective
and implied financial sovereign that oversees each transaction. Who or what "evaluates" derivatives in the absence of a market?

The Derivative Subject; or, A Specter Is Haunting Markets

Today, we must ask ourselves whether Wall Street's young rocket scientists, like the old

Tom Lehrer song, have concluded that "I just shot them up, I don't care where they come

down, that is not my department, says Werner Von Braun."

Representative Edward Markey

In 1992, faced with recent high-profile derivatives losses, the House Subcommittee on Telecommunications and Finance inquired into the derivatives market. What new risks does the market pose? What regulations can and ought the Federal Government enact to avoid "a chain reaction of losses and defaults that drain liquidity from our financial markets"?[29] The House and its expert witnesses, in their figuration of exotic derivatives as "recombinant financial DNA," mark out the peculiar character of financial subject installed by such "financial engineering."

For the threat of derivatives, from the perspective of *both* the regulation-friendly Congressional speakers and the experts who filed into the committee room poised to resist such regulation, is the derivative's distance from an origin. Like the recombinant DNA techniques with which it is rhetorically rendered, derivatives thwart any attempt to map out a "natural" relationship between an original—organism or market—and its simulacrum. Indeed, it is precisely in the space between the market and its various models that the derivative lives. Consider the following exchange concerning the evaluation of derivatives between then Representative Bill Richardson of New Mexico and Peter Vinella, a senior consultant at Smith Barney Shearson, and an expert on derivatives.

> MR. RICHARDSON: Mr. Vinella, given the uncertainies and assumptions surrounding marking to a model, how reliable are the values derived from such a process?
>
> MR. VINELLA: Well, in some cases it is the only chance you have of coming up with a price. As the issues become more specific to an individual investor's needs, they become less traded on the secondary market, therefore, mathematical models are really the only hope of coming up with a good idea of where the security should be priced.[30]

To parse the rhetorical economy of these transactions: Exotic derivatives, financial instruments with near zero liquidity—they cannot be resold on any market; in the words of one expert, "The Market is you"—provoke a crisis of evaluation. With no external market in which to be bought and sold, evaluated, the derivative floats without reference to any "price." How are firms to disclose and report the risk and value of derivatives? Since the very structure of capital in its accounting demands precisely such a disclosure—one must, in some sense, know *what* has been bought, what is owned—something other than the market produces the required evaluation "effect." This evaluation is effected through "marking to model," computer simulations that evaluate, if not determine the "price" in a kind of Sim market.

Hence the stories of loss associated with the derivative, ultimately the loss of reference and the production of anticipation. For the derivative can only have an anticipated value, as the difference between the simulation of a market and a market is a difference that cannot be bought off, an unpredictable algorithmically complex difference that will not go away, at least not right now. Therefore something other than contemporary human subjectivity or markets underwrites or *advances* the derivative, perhaps a human that becomes the financial subject only retroactively, perhaps only after its death; the derivative casts its shadow from the future.

> In the stock and bond markets, the public dissemination of price quotes and transaction information allows all market participants to see clearly the current price of a security. But when it comes to derivatives, for now we see through a glass darkly. So who knows what prices and risks lurk in the OTC derivatives market? Often, only the "Shadow" knows.[31]

Thus the distance of the derivative from a "price" is cast as a loss of knowledge and of vision, a shadow cast from a lurking presence in the future, a presence that, "for now," projects an unclear specter. The structural impossibility of evaluating the derivative in the present—for it can be modeled, and not known, and the precise risk and productivity of the derivative resides in this difference—raises the question of who or what will "know" the value of a derivative in the future.[32] For Fredric Jameson, the fundamental inaccessibility of this knowledge is characteristic of late capitalism in general:

> These new and enormous global realities are inaccessible to any individual subject or consciousness—not even to Hegel, let alone Cecil Rhodes or Queen Victoria—which is to say that those fundamental realities are somehow ultimately unrepresentable or, to use the Althusserian phrase, are something like an absent cause, one that can never emerge into the presence of perception.[33]

But rather than following Jameson here in his formulation of an "absent cause" that fuels global capital, the exotic derivative seems to demand a mapping of a *cause from the future*, not absent, but to come. For while the reality of exotic derivatives is "unrepresentable" in the present (the "presence of perception"), they nonetheless yield an anticipation of a future perception, a future evaluation, a future reading. If exotic derivatives resist representation in the present, it is only because they are, like the cryonic body, residing in the future.

While the discourse of expertise that accompanies derivatives also focuses on this narrative of loss, a narrative in which the present is the site at which a network of humans and machines fail to evaluate the exotic derivative, we might instead notice what it is that derivatives produce in the present. Peter Vinella, a senior consultant with Smith Barney Shearson, spoke about the use of derivatives as a financial tool and not merely a hedge:

> One of the examples is McDonald's going into Japan...In order to fund that, they went into the derivatives market. They issued corporate debt here and used derivatives to translate that into the Japanese operation...Just looking at the derivatives contract by itself would be meaningless. You wouldn't understand why McDonald's came up with this complicated structure. If you stress tested it by itself, you wouldn't understand that underneath it they are also selling hamburgers, which is the other side of the trade.[34]

"Meaningless" in the present, the derivative contract described above nonetheless yields something now: anticipated hamburgers, Big Macs to come. As a speech act, then, exotic derivatives produce not "meaning" but anticipation, an effect in the present, a becoming-burger.

This economy of anticipation also seems to be provoked by the exposure to concepts of cryonics. In the preface to Robert Ettinger's manifesto for cryonics, *The Prospect of Immortality*, Jean Rostand sees a business opportunity:

> While reading this book, I was reminded of the Belgian businessman who in the early days of World War II heard rumors about the possibility of atomic fission. He ordered a large supply of uranium from the Congo and sent it to warehouses near New York just in time for the atomic bomb project. I must confess that were I interested in business speculation, I should be busily stockpiling equipment needed for Mr. Ettinger's project.[35]

In the present, both derivatives and cryonics effect anticipation, anticipation of a future that will transform both the cryonics contract and the derivative algorithm into meaningful speech acts. Rostand's confession tells the truth about cryonics: the future that is bundled with cryonics is primarily a site of anticipation, a moment for which one should, in the present, "busily" prepare. This is a common enough figuration; the future, death-oriented character of the modern subject is not news.

But the notion of the future built into cryonics and exotic derivatives marks out a peculiar difference in this notion of preparation: the unverifiable nature of the future is turned into a positivity, a condition that enables rather than thwarts the production of new knowledges and capital formations. Both cryonics and exotic derivatives thrive as capital formations that are *beyond the market.* Based on shadows, "rumors" of the future, both the cryonics and the exotic derivative contract are unverifiable in the present. As such, they are not only sites for a cause from the future—an anticipated cause, one we busily prepare for—but also sites disciplined by the future.[36]

Disciplined by the Future

Whereas Foucault has emphasized the panoptic *spaces* of discipline, the visible and yet unverifiable operation of the Panopticon, a machine that mobilized uncertainty in the production of criminal subjectivity, contemporary global capital seems to be fueled by visible and yet unverifiable inputs from the future—a future body of cryonics, the future capital of the exotic derivative. As a kind of Brennschluss of subjectivity—*I just shot them up, I don't care where they come down, that is not my department, says Werner Von Braun*—both cryonics and exotic derivatives are constituted out of, driven by, the uncertain future that inheres in each. Indeed, the cryonic body *is* the thought of a subject for whom the uncertain future of capital is articulable. Cryonics is not merely the export of a subject into the future; it is the evaluator and the body of a capital that is yet to come.

Thus both exotic derivatives and cryonic bodies exchange an economy of reference for an engine of anticipation. Neither refer, in the present, to anything; they are contracts with the future. The status of these contracts is, of course, contingent in the present, as undecidable as the character of Disney's body. Contemporary global capital appears to need this difference, a difference between the present and the future that is undecidable and yet endlessly evaluated.

This notion of the future as an uncertain positivity, an entity that can be mobilized and encountered in the present, is neither the Laplacian dream of transcendence nor an eruption of prophecy. For both cryonics and exotic derivatives, the core of undecidability that inheres in each contract is a *natural* consequence of its hybrid status. A cryptographic body, generated out of the enmeshed ecologies of metaphor in the present and nanotechnology in the future, is tied to the finitudes as well as the possibilities of the very computers that enable it. For computers, too, are haunted by contingency:

> If we knew what the result of a computer program would be, there would be little point in writing or running it. In fact, it's a fundamental theorem of computer theory that in the general case, there's no way to tell what a program will do, other than to run it and see what happens.(CryoNet, message 5310)

The dual dice throws of cryonics and exotic derivatives are thus constituted out of the material constraints of computation, disciplines of the machine whose status only becomes clear in the future. Disciplined by the future, the present anticipates but suspends judgment. Mike Darwin, the cryonics technician discussed above, inhabits this contingency, recognizing the signs of discipline on bodies in the present, but unable to evaluate them. Speaking of the effects of current cryonic techniques on bodies, Darwin eschews discussion of abstract models of human memory:

> Since we are apparently grinding brains into chopped steak (I won't go so far to say hamburger, and get everybody upset) with existing techniques I see little point in spending my time on evaluating memory in a model which does not have much to do with what we are *really* doing to human cryopreservation patients. (CryoNet, message 4117)

The real, rather than "apparent" effect of the present will perhaps only, become visible in a future in which the becoming of exotic derivatives will also materialize in the form of new infrastructure, new capital, new burgers. In the meantime, the difference between hamburger and cryonic bodies is, at the moment, in the end, uncertain. Suspended.

F   O   U   R

"Give Me a Body, Then":
Corporeal Time-Images

The brain has lost its Euclidean coordinates, and now emits other signs.

Gilles Deleuze

Faxing Lazarus

SO WHOSE car was this anyway? It was like waking up in a strange room, those few moments before you figure out where you are, where the bathroom is, the time. A friend lent me his car to drive the hour or so—distance always being imploded into hours in "The Southland"—from Irvine to Rancho Cucamonga, and for a blink, a long one, I forgot that I was encased in somebody else's appliance, hurtling past Disneyland. The cars ahead slowed, and I fumbled around with the map, wondering if I had already missed one of the numbers I was supposedly searching for. The red tendrils that traverse the map reminded me of visuals of a synapse I had seen when I was a freshman in college, which is precisely my problem with maps: they always make me think of something, someplace, else. Paul, the friend with the car, had written me out a recipe for getting there: 73 north to 55 (forks right) to 5 to 57 to 10 east. This was his way of helping me avoid getting lost or arrested.

I had spoken four days earlier with Mike Darwin, head of 21st Century Medicine, subcontractor for the cryonics organization CryoCare. CryoCare was a new upstart cryonics care provider, located close to the Ontario, California airport for high-velocity body delivery and cooldown, perfusion. You were two flights and a short pause of a layover from almost anywhere, so whatever forces carry bodies all over the world in order to freeze them had an eddy, a whirlpool of anticipation and frenzied movement, in Rancho Cucamonga. Darwin, though, was no novice. Exact numbers are difficult to come by in the world of cryonics—we will only have them with revival, retroactively, after the Singularity and nanotechnology reconstruct us atom by atom—but Mike had frozen scores of humans, and plenty of dogs, including Lazarus, a dog I was about to meet.

I was speeding, doing eighty-five even though I had no driver's license, because I was stupid. I was stupid because I had fallen asleep in the library reading a book about comas—*The Catastrophe of Coma*—as a way of trying to come to terms with Deleuze's understanding of the new cinema, "A disturbed brain-death or a new brain which would be at once the screen, the film stock and the camera, each time membrane of the inside and the outside" and left late for my 5:00 P.M. appointment with Darwin.[1] I *had* to go fast; I had, after all, justified the whole trip to California in terms of the alleged interview I was to make with some cryonicists. Here I was with no tape recorder, no idea of questions, and increasingly late. What was I doing?

The mountain, whose name I don't know, had a dusting of snow, and there was no smog as I drove through Chino, a town with long, wide streets, asphalt ribbons peeling off the mountain like so much skin. Often, the streets flowed here with thick, biting smog, making your eyes burn until you look and feel stoned from some of that Humboldt bud the hippies bring down on Interstate 5 and sell in little numismatic bags in Venice. A recent rain and a wind had freshened the air, and for some reason I was reassured. It didn't look like Southern California.

Haven Street leads to Civic Center, and Darwin instructed me to start counting fire hydrants as I turned the corner, since his was between the sixth and the seventh on the street. Had I missed one? Where's the first one? Everywhere I looked, fire hydrants. I'm in the middle of a fire hydrant forest. Must get out of this navigational vertigo. I saw a strip of buildings with humans, and I searched for a phone but was denied. "They're turned off," a woman standing by the door of an investment company said, "they're off." I didn't think about it at the time, but since when are phones something that get turned *off*?

Yes, yes, "Bad omen!" you're saying, as I find myself sealed off in a concrete extrusion of global capital, grown in a petri dish of low taxes and airport proximity. This is a place that exists to move practices and knowledges *elsewhere*, and yet the phone, that great vector of global movement—"Honey, I shrunk the world..."—was off.

Whose car was this anyway?

I saw the promised orange number in the right-hand corner of the building that Darwin had foretold, so I breathed some relief and pulled in next to a white four-wheel drive. A Bronco Blanco, as they said in the O.J. Simpson trial. Cryonicists *would* have four-wheel drives, I thought. They would always need the reliable means of movement that comes with the sudden freezing of bodies and their constant surveillance and care.

It was a room much like the one at Gentle Dental, a low-budget dentist I had once submitted myself to while at MIT. I wanted to press my nose against the glass, see what was beyond the first room, and then a small, bearded man spotted me and heaved the door toward him.

"Are you Mike?" I asked, somewhat astonished that this person actually existed in the midst of the hydrants, asphalt, and long, low buildings that hugged the ground in search of safety from the sudden seismic messages of Southern California. "No, no," the man with the wispy beard said, "no." He gestured toward a group of rooms that unfurled from the main hallway. "No."

"We already paid this bill. I'm not paying for electron microscopy twice." A tall man with thick, well-cropped gray hair and wire-frame glasses was surrounded by documents, computers, desks gone to office entropy.

"Here's the person that you were waiting for, Mike."

Mike Darwin pumped my hand with a real estate grip as we drifted back toward the front room, floating toward the memories of bad dentistry. I checked a cavity with my tongue.

The dog, it seemed, had been frozen. Its companion, a friendly cat named Sam, sprouted a twelve-pin jack from its furry, scabby head, so that it might be more readable, so that its EEG could be checked at any time, a time in which the object was to pause time, to freeze a cat or a dog and bring it "up," as if nothing had happened. Darwin scratched the cat's head. "It itches," he said.

At this point the soundtrack of the image sculpted here becomes dense with barking. Dogs, lots of them, told Darwin that they were tired of waiting, waiting for food, water, Frisbees, biscuits. All this waiting.

Darwin told me about Lazarus. He had been "down," as Darwin put it, for sixteen minutes. Slowly, his canine body had cooled down to well below zero, until he was, semiotically speaking, dead. No sign of vitality. Just as slowly, after an intermezzo, Lazarus was warmed, until all the signs of life returned. Now he had the run of the lab.

"Our goal," Darwin said, "Our goal is to get to the point where we can cool him down, pack him in ice, put him in a casket, and fly him off to Florida. Then we can offload him to Disneyworld, all the media there, and bring him back."

"You want to fax the dog," I said.

"Yeah, fax the dog. Or at least the dog's life. To fax it we need to get to six hours, not just sixteen minutes, but I'm a patient man."

A Fold, Grafting the Future

... the rise of situations to which one can no longer react, of environments with which there are now only chance relations ...

Gilles Deleuze, *Cinema 2:*
The Time–Image

Gilles Deleuze, in his multivolume work devoted to the cinema, locates a "mutation" within cinema, "new signs invading the screen" after the Second World War.[2] Deleuze maps this transformation of the cinematic field not in terms of a shift in the relative values of a genre, the new technologies associated with the special effect, nor to an understanding of the shifting relations of cinema to its mode of production. Rather, Deleuze locates a shift in the concepts through which cinema renders time and movement. The movement-image, for Deleuze, rendered motion through an assemblage of image/cut/image, one image following another. It did so via the "association of images," one image after another, each one crossing the void of movement. Thus in the movement-image, the assemblage of the film apparatus and the brain yielded movement through the *disjunction* of individual frames and shots. Movement, then, was not given "in" the image, but was instead composed out of the gaps and voids of the cinematic space, as if the image were moving itself, with periodic moments of rest. These gaps or cuts were not breaks or additions to the frame or immobile sections that made up cinema. Instead, Deleuze locates a sequence of shots as an assemblage, frames in motion through the production of univocal, unified space that is traversed by an irreducible movement.

> The space covered is divisible, indeed infinitely divisible, whilst movement is indivisible, or cannot be divided without changing qualitatively each time it is divided.[3]

Why is movement itself indivisible? As interruptions of a plane or space that is smooth, divisible yet full of connection, cuts cannot themselves be cut, at least not without a change in the character or quality of the cut, a point of transition that would constitute a different fold of cinematic space. A classical filmic sequence, then, rhetorically rendered movement through an assemblage of cuts, voids that garnered their effects through their contiguity with images.

Topologically speaking, then, the movement-image emerged out of folded images, pockets of movement whose interiority continually referenced the image or frame that was put into motion. The cut that connects, the movement-image was riddled with holes but nonetheless formed a chain of commensurables, movements of the same kind that traversed a stable space that would reference them.

With the time-image, the cinematic sign changes in character. No longer a series, where each image is commensurable with the last on a unified, divisible plane, the time-image is constituted out of the interstice itself, an intermezzo that begins to have an importance in itself.

> The modern image initiates the reign of "incommensurables" or irrational cuts: this is to say that the cut no longer forms part of one or the other image, of one or the other sequence that it separates and divides.[4]

This transformation of the cinematic sign—from a regime that renders formal linkages of images to one that comports a continual break, "the irrational cut" that is grafted *as a cut* and not a becoming-linked—reverberates with the entire assemblage of subjects and machines that composes cinema. If early cinema and the production of the movement-image demonstrated the "impower of thought"—with the brain's "filling in" of movement through the flicker of the image, one could not look away—the time-image sometimes provokes the impower or even the ends of subjectivity, a subjectivity that is only about to occur, an algorithmic subjectivity whose instantiation is ongoing and subject to sudden transformation. In his discussion of the "media effect" produced by German filmmaker Hans-Jürgen Syberberg, Deleuze notes that the "interstice" between the visual and the sonic renders not subjects entranced to cinematic Caligari cabinets but rather maps a landscape of informational complexity:

> The disjunction, the division of the visual and the sound, will be specifically entrusted with experiencing this *complexity* of informational space. This goes beyond the psychological individual just as it makes a whole impossible: a non-totalizable complexity, "non representable by a single individual," and finds its representation only in the automaton.[5]

This invasion of the screen by new signs, then, marks more than the emergence of a new style or possibility in cinema; it maps a transformation of the effects of representations and the subjects that would bear them. While the movement-image, with its kernel of transformation buried "between" each image, mimed and perhaps periodically constituted a subject of interiority whose depths were unrepresentable but potent, true, and secret, the time-image, with its incessant interruptions of itself, invests the intermezzo of the image with a positivity and complexity that is "beyond" the subject. Not just "between" images as an interval that exceeds the inside and outside of any image, the time-image is in some sense "between" subjects, neither inside nor outside but resonating in a space of informational complexity that resists any interiority. No longer referenced to an image that would engulf it and force it to refer to a stable, commensurable space of movement, the time-image dwells less as a cut that forms an outside for the interiority of an image than in the Möbius topology of a cut that is neither inside nor outside but in a differential fluctuation. Here the cinema is less a parade of images to the sovereign spectator, riddled with flickers, than a relentless production of an intermezzo, the image of time in its positivity, composed of the entire assemblage of spectator, film stock, screen, and camera, an assemblage which can only be endured as an actualization:

> Everything can be used as a screen, the body of the protagonist or even the bodies of the spectators; everything can replace the film stock, in a virtual film that now only goes on in the head, behind the pupils.... A disturbed brain-death or a new brain which would be at once the screen, the film stock and the camera, each time a membrane of the outside and the inside?[6]

Hence, for Deleuze the time-image fosters a body that lives through this intermezzo, is the intermezzo, always waiting:

> The body is never in the present, it contains the before and the after, tiredness and waiting. Tiredness and waiting, even despair are the attitudes of the body.[7]

Thus the "interstice," the interval that constitutes the time-image, marks an anticipatory period of waiting, but it is not a preparation for a link that is to come. Instead, the fissures comported by the time-image are waiting only on themselves, pulsing with a contingency that would traverse them, shattering each gap anew with another interval, another cut whose link can only be made of contingency. Neither before nor after but becoming, the interval set free in this manner affirms time as an exteriority, an unprecedented contingency:

> We no longer believe in a whole as interiority of thought—even an open one; we believe in a force from the outside which hollows itself out, grabs us and attracts the inside. We no longer believe in an association of images—even crossing voids; we believe in breaks which take on an absolute value and subordinate all association.[8]

The efficacy of Deleuze's claims for this cinematic shift could and should be interrogated in terms of their capacity to map problems and tensions in the history of cinema, and yield encounters with images that provoke events other than the habitual recuperation of subjectivity. But the emergence of this intermezzo, this fracture, this break, as a semiotic operation of the cinema is involved in more than cinema.

> A theory of cinema is not "about" cinema, but about the concepts that cinema gives rise to and which are themselves related to other concepts corresponding to other practices, . . . It is at the level of the interference of many practices that things happen, beings, images, concepts, all kinds of events. The theory of cinema does not bear on the cinema, but on the concepts of the cinema, which are no less practical, effective, or existent than cinema itself.[9]

Deleuze noted above that it was only with the figure of the "automaton" that the excessive topology of informational complexity could be rendered. If the time-image, as a practical concept that operates in and through cinema, is caught up in the interference of other concepts and their practices, it is perhaps "informatics," with its relentless spawning of the intermezzo space of networks, with which this concept can be mapped.

Becoming Corporeal: From "That's All There Is" to What Will It Become?

This story of Lazarus narrates a transformation in the concepts and sciences of "life," transformations that transport the bodies of organisms as materially as a 747 to Orlando. More than a becoming-machine of the organism, this retooling or "refiguring" of life provokes double takes on the becoming-lively of the machine. The recipe that makes it possible to plan for a future "faxing" of an organism, then, begins with the command: transform into information.

This transformation of organisms—and their deaths—into information, of course, has all the hallmarks of a narrative of nostalgia. Once upon a time, so the story goes, biology concerned itself with the dynamic vitality of organisms, an interiority of struggle that offered a density and thickness to the object of the scientific gaze, a density that is all but evacuated by this rendering of organisms into code. Thus molecular biologist Walter Gilbert narrates this shift in the articulation of life as a movement from organisms to databases:

> In the current paradigm.... The "correct" approach is to identify a gene by some direct experimental procedure—determined by some property of its product or otherwise related to its phenotype—to clone it, to sequence it, to make its product.... The new paradigm, now emerging, is that all the "genes" will be known (in the sense of being resident in databases available electronically).[10]

This postvital paradigm, then, no longer looks to the body of the organism, its phenotype, for the production of truth. Rather, the new knowledges of life will, Gilbert argues, proceed through the relations of biology's "new reagent," information. The remarkable ascent of the molecular understanding of life—Watson's "to understand what life is, we must know how genes act"—emerges through the eclipse of the organism, the implosion of life into "information." So too does that other threshold, death, become a problem of an informatic kind. Ralph Merkle, cryonicist and cryptographer of postvitality, writes of the new distinction wrought on death by information:

> This true and final death is caused by loss of information, the information about where things should go. If we could describe what things should look like, then we could (with fine enough tools, tools that would literally let us rearrange the molecular structure) put things right. If we can't describe what things should look like, then the patient is beyond help. Because the fundamental problem is the loss of information, this has been called information theoretic death. Information theoretic death, unlike today's "clinical death," is a true and absolute death from which there can be no recovery. If information theoretic death occurs then we can only mourn the loss.[11]

The concept of information theoretic death, then, emerges through a recursive blur of vitality: the becoming-lively of machines, the becoming-machine of evolution. And Merkle, too, follows the rhetorics of information unto nostalgia and death. If the loci of life have shifted from organisms to databases, there is nonetheless a paradoxical remainder of vitality's story and its nostalgias: "we can only mourn the loss." This sudden evacuation of possibility provoked by information theoretic death indicates a limit of algorithmic processes: that non sequitur of physics, irreversibility.

The specter Merkle anticipates mourning is linked to an apparent difference between physical and biological systems: living systems are continually open to surprise, irreversible changes that can never be "put right," if only because biological "things" are less objects than ongoing transformations.[12]

If irreversibility emerges as a terrifying problem rather than a remarkable attribute of living systems, it does so at least in part under the influence of a sequential model of life. Under this regime, information—the sequenced genome, the diagram of development inscribed by a laser and the ablation of cells, and the complete "neural circuit" of, say, a worm—becomes represented primarily as a "sequence," genetic "actors" understood in terms of the next element in a sequence. With the sequenced genome, the genetic "actors" are understood primarily in terms of what gene product a given sequence gives rise to, cells are traced within a trajectory of development, the "next cell," and neural operations are mapped via the "behavior"—the next action they yield. The "frames" that compose each sequence are thus comported in terms of their "movement," a movement that emerges out of the traversal of a infinitesimally tiny gap, the difference between a gene and its gene product, one cell and its descendent, and so on. As such, this visualization of living systems proceeds through an operation of the movement-image, a sequence of images whose effects emerge out of the association between one moment and the next, moments themselves traversed by an invisibly actualized difference, life.

The object of molecular biological knowledge rendered here is thus sought within the frame—the genetic actors—and not in the interval between frames, the actualization of transformation themselves. These transformations—the networks of transcription and translation that "read" the "reading frame" of DNA, and the operations of protein folding that yield three dimensional structures from sheets of amino acids—become strangely unpredictable supplements to life's "actor," the code. Like the computer algorithms with which they are rhetorically linked, though, the instantiation of almost any algorithm involves an irreducible event of becoming: just run it. Hence what is imaged in this high-resolution map of "what life is" is not the differentiating outburst of evolution known as "life," but the movement of gene–to–gene product, neural circuit to behavior, movement itself. Life and corporeality, the transparent body, any organism whatsoever forms the divisible and invisible substrate upon which such movement transpires, a self-moving image called "DNA."[13]

But if this remarkable resolution frames the organism as tedious repetition of theories of motility, perhaps Lazarus, in his long pause, is a rendering of organisms in terms of their irreversible becoming, what they are capable of. It may seem that this experiment in cryonics is a massive investment of discipline into the frame, a dog actor who would be recuperated at the end of long journey across a void, a kind of postvital Lassie come home. As a technology of immortality, nothing would seem to comprise a more sutured interiority, that wetware of the movement-image, than the desire to return from the infinite gap of death.

But in "faxing" Lazarus, Darwin comports him as an informational construct that is put on hold. The entire operation of cooldown, perfusion, transport, and revival is devoted to a production of what will have been an intermezzo, a space between life and death, inside and outside of time. This intermezzo, of course, only gets produced as a moment "between" moments of Lazarus, but it is this pause—sixteen minutes or six hours—that garners importance "in itself." The space between life that makes possible a "faxing" of an organism—a dog reduced to a readout of information, that readout of information is transformed back into a dog—is oriented not toward a statement of what is, but involves an experiment into life's capacities. Less about an actor than about transformation, a transformation into information, this cryonics experiment tinkers with life rather than describes it. Neither inside nor outside life, Lazarus forms a Möbius body. Neither inside time—the dog is precisely clocked—nor outside it—it continually encounters contingency—a loss of power, a loss of life—Lazarus indexes the time-image: "Time-images are not things happening in time, but new forms of coexistence, ordering, transformation."[14]

And yet "all the media will be there," and "there" will be Disney-*world* and not Disney*land*. How long will it take to notice? When will this difference occur, and how will its difference be imaged, imagined? At once, this intermezzo is territorialized into movement—the second part of our Lazarus algorithm would read: transport to Orlando. It is as if Lazarus's pause, spreading from the middle, can only bear representation, or perhaps narration, as a journey in space. But the precise character of the journey provokes not movement from one image of America to another, but simulacra repeating, echoing each other.[15] This evacuation of movement foregrounds the positivity and materiality of time, the complex intermezzo of frenzy, aircraft, video, and terrible, extraordinary cold. A six-hour trip to a future of itself, the journey from Disneyland to Disneyworld becomes less the traversal of a continent and more the repetition—or a fold—of an entire field of automata, those machinic characters that can bear, Deleuze claims, the representation of informational complexity. It is, perhaps, as if Lazarus blinked an extraordinarily slow blink in Pirates of the Caribbean, immobilized by the cold. Lazarus does not so much move from Disneyland to Disneyworld as he erupts in Disneyland and Disneyworld, an assemblage of simulacra networked with the frozen flesh of Lazarus. Moving so fast that he is motionless, Lazarus will have embarked on time travel, a journey toward Disneyland of the near future, six hours from "the present." Here the difference of the future, as in all time-travel scenarios, is too great to be born by representation; the future "traveled" to must be only slightly divergent from the present, a divergence much less than the geographical distance between Anaheim and Orlando, and incalculably more than nothing.

The importance of the irrational cut, the time-image, as a way of mapping the contours of contemporary corporeality, resides less in its truth or fidelity than in its complicity with *futures*. For the intermezzo, in its grafting of the imperceptible if not absent organism to the future, poses the question, the problem, and the ecstasies of becomings proper to the body, what bodies are capable of, and therefore what they may become, then. "Give me a body then"[16] is Deleuze's formula for a philosophical reversal, one that takes the body not as a "resistance" to thought, but as that which provokes thought, puts it into flight, forces us to think. Biological bodies, then, even as they are networked and even imploded by information, become occasions not just for the astonishing connections of a so-called biological multiplicity and its alterities. They also reek of rhizomatic entanglements with capital, the massive and material exteriority of this technological déjà vu, hurtling toward Disneyland. Whose car was this, anyway?

F   I   V   E

"Remains to Be Seen":
A Self-Extracting Amalgam

Waiting and suspense are essential characteristics of the masochistic experience...

Gilles Deleuze, *Coldness and Cruelty*

October 12, 1998

IT SEEMED likely that I would begin working on the interview in the near future. Sometime soon. A Friday would be best. At some blurred point in the past, Thursdays had become consumed, gnawed away by teaching and the anticipation of a weekend. Friday was like a relief from the continual imaging of a future that wasn't so imminent, instantaneous. In a few days, I would have the capacity to be worked over by the delays and relays of writing.

Without a doubt, this is the source of my flight, this uncanny overtaking, even possession, that seems integral to ecologies of writing. Some perverse fidelity to my research demands that I be transformed by it. This is not exactly something I can do on purpose.

October 15, 1998

I have changed my mind about the interview. It would make more sense to do it by phone. All of my hopes of frequent flyer miles evaporate. I'll make some coffee, pick up the phone.

Robert Ettinger is the founder of the Cryonics Institute in Clinton Township, Michigan. For 28,000 dollars (US), Ettinger's institute will freeze and store your legally dead body in preparation for future revival. CI is one of several players in this relatively new field of customer service. Alcor, now in Arizona but formerly of California, is the largest cryonics care provider,[1] while CryoCare, of Rancho Cucamonga, California, is the most recent upstart to compete for the cryonics patient. The numbers involved with cryonics are not huge—the number of bodies and/or heads frozen is probably in the hundreds rather than thousands—but as with the value of Amazon.com shares, the business of evaluating the relative worth of cryonics organizations is predicated on unprecedented future success.

If origins matter—and in the practice of cryonics, this is a matter of some dispute—then Ettinger has the credentials. In 1962 Ettinger, a college physics and math instructor, published *The Prospect of Immortality*, a text that proved to be a recipe for generations of cryonics patients.[2] Ettinger offered a vision of the future as one of plenitude, plasticity, and wealth. While the present was to be enjoyed at all costs, cryonics offered a future not just of immortality, but of differentiation. Freezing was a practice of becoming—even English professors would be transformed, as human culture would undergo a change in kind associated with immortality:

> It is true that we still read Beowulf, and the Iliad, and Hamlet, and many scholars blithely assume that these and similar works will remain in our culture forever. But in the last thirty or forty thousand years, the supposed tenure of modern man on earth, cultural changes have been relatively small, and biological changes virtually nil. In the next few centuries, the changes will be incomparably greater...I am convinced that in a few hundred years the words of Shakespeare, for example, will interest us no more than the grunting of swine in a wallow. (Shakespearean scholars, along with censors, snuff grinders, and wig makers, will have to find new, perhaps unimaginable occupations.)[3]

October 24, 1998

I began researching cryonics more than five years ago. You'd think I would have signed up by now. I had run into some stuff on cryonics in a newsgroup on nanotechnology that I regularly read, sci.nano.[4] Somehow the creation of self-reproducing molecular machines was involved in the forecast for cryonic revival. The future capacity of these critters to assemble copies of everything promised much for those who were, as Ettinger put it, on this side of the freezer. Ralph Merkle, a nanotechnology researcher at Xerox PARC and cameo brand name in Neal Stephenson's *The Diamond Age*,[5] offered the cryonics community the notion that the frozen cryonic corpse was a scrambled text waiting for decryption. Nanotechnological appliances of the future were the likely exegetes. Once I realized that contemporary cryonicists conceptualized bodies as ensembles of information that could be decoded, I was hooked. Cryonics, I realized, was positively mainstream. Cryonicists were, to be sure, the last gasp of an organismic biology—they thought the body mattered enough to freeze it!—but they nonetheless were sensible enough to accord with contemporary molecular biology, which tells us that living systems are just interesting configurations of information.[6] I wondered if they were interesting enough to experiment on.

October 24, 1998

I had assumed, at first, that Ettinger was dead. This isn't really a fair assumption, since it's only 1998, thirty-six years after Ettinger's first book. I, after all, through some rather elaborate Rube Goldberg devices, have survived since 1963, so why assume otherwise for Ettinger?

Maybe it was the spooky character of the research. I'll admit to being a bit panicked by my early readings in cryonics. For it is one thing for Richard Dawkins or *Gattaca* to *describe* the lives of human beings as networks of information. But cryonicists were devoted less to description than to practice. They *treated* human bodies as elaborate memory structures, and devoted much of their lives to the preparation and invention of these treatments. They took the news of the informatic evolution of humans to mean: To Be Continued. In this scenario—although these are not the terms of the cryonicists—it's easy to feel a little bit haunted.

Merkle's "Cryonics, Cryptography, and Maximum Likelihood Estimation," for example, suggests that our contemporary or "clinical" declarations of death are premature. It ain't over, Merkle suggests, until the very memories of the deceased—corporeal and collective[7]—are scrambled, a state, as discussed in the previous chapter, Merkle characterizes as "information theoretic death."

This persistence in the course of physical death—the tendency for the properly treated body to memorialize its own life, to testify to its own disease—renders future information even when it ceases to be capable of any ongoing self-maintenance or autopoiesis. In short, it becomes a sign for the future, a recipe for "put[ting] things right."

But for whom does this dead body become a sign? The project of maintaining the clinically dead body is now outsourced to others. If the ongoing process of pushing entropy around is too much for the individual, a clinically dead individual who excels only at nothing, could not a community—of cryonicists—collectively manage the entropy?

This subcontracting makes a bit unclear just who is doing the *living* in these cryonic ecologies. Imagine working for corporations whose workforce and profits are tied to such continual maintenance, to live for others who can't do it on their own. Perhaps this will be one of those "unimaginable occupations" about which Ettinger prophesied.

No wonder I thought he was dead.

October 26, 1998

I had once e-mailed Ettinger with praise for *Prospects*. I can't find it now—that was three or four hard disk crashes ago. When I had first seen ettinger@aol.com on a message from CryoNet, it was like a brush with greatness—could it really be the cryonics pioneer? And of course there's always something slightly uncanny about meeting the author of an impressive book—it's difficult to avoid seeing the author as the double of her work. Novelist William Gaddis nicely reversed the polarity of this description when he diagnosed the "passion" for meeting writers and artists to supplement the encounter with the work. "What do they expect? What is there left of him when he's done his work? What's any artist but the dregs of her work?"[8]

Subject: Interview
Date: Mon, 26 Oct 1998 12:01:14 -0500
From: Richard Doyle <mobius@psu.edu>
Organization: Hybrids Unlimited
To: Ettinger@aol.com

Dr. Ettinger:

You may recall that I wrote you about a year ago inquiring into your current work. I am an assistant professor of English at Penn State University, and I am interested in the relationships between science fiction and scientific research. I have written a book, "On Beyond Living," on Stanford University Press, that details the historical transformations in our understandings of living systems, focusing on the history of molecular biology. I have been researching cryonics for several years now, and am interested in talking with you about your current work, your reflections on the history of cryonics and the future of the human body. I would like to interview you sometime in the next month if that would be possible. Would you be agreeable to such a chat? I would very much appreciate any time you could give me.

Best Wishes,
Richard Doyle

December 31, 1998

9:45 AM

In the beginning was the busy signal. Along with a hiss that announced the low cost of my recording and playback apparatus, I heard a refrain that told me to wait. All circuits are busy. Try again later.

Already, there had been much waiting. A light, chalky snow dusted the early morning traffic of the college town where I live, enough of a coating to enjoy a periodic fishtail as I negotiated the Wal-Mart parking lot, a wide, flat pool of asphalt bordered by a bagel franchise and a video store. There are two Wal-Marts in State College, Pennsylvania—the first had, for some reason, proven to be the highest grossing store in the country, so a copy of it was installed north of town, near the Barnes and Noble/Starbucks that had sprouted in the proximity of a six-screen multiplex. The original Wal-Mart was closer to campus, where I sometimes work, so I pointed the truck between McDonalds and Bi-Lo Foods, the wipers on low to clear away the meager but persistent snowfall.

The aisles, of course, were laid out differently in each Wal-Mart, and I could never, fortunately, remember which was which. Someone greeted me at the door, rendering customer service. "Happy New Year."

The Magnavox/Phillips AQ6581 stereo radio cassette player has Dynamic Bass Boost (DBB) and automatic stop.

9:17 AM

I remember to buy some batteries. My usual strategy of cannibalizing AA cells from a VCR remote won't do—it's a forty-five-minute interview, if I remember correctly. I buy the two-pack.

I am sliding, sliding my Mellon Bank card through a beveled slot—it's called "swiping"—but the display continually insists that the card is "unreadable." And here I was just paid. The guy who is working the counter—mid-thirties, bearded, the thick body of an ex-football player—blames it on the cold. "The magnetic stripe gets brittle." He tries warming it in his palms and reswiping, but no dice. There are sixteen dollars in my wallet and I unfold and smooth them onto the counter while the tape player package and the batteries in their tester modules are dragged over the red laser. "Happy New Year," I offer. "Party on," I am reminded. "Crank it up."

9:35 AM

Due to budget efficiencies in the context of a billion-dollar fundraising campaign at my university, the heat is off in my office. I have attacked the tape player packaging with some old scissors, and I followed the diagram inscribed on the battery compartment, clicking and springing the cells into place. The package claims that they are good until March 2003, probably longer if I keep them in the fridge. Headphones on, I "play."

9:36 AM

This silence is worrisome. I don't even get any hiss. No LED announces the use of power. I go into troubleshooting mode.

9:40 AM

After checking the headphone connection—which was loose—and trying different strategies of battery placement, I realize that I have forgotten about the tape. Hiding in my back pocket, it revealed itself only when I sat down in my ancient office chair in a clever, almost Zen effort to relax and locate the flaw in my troubleshooting. I think I heard something snap.

9:44 AM

The tape is extracted from its case and found to be unharmed. A short segment of tape has strayed from its housing, an errant fold that is easily removed by turning one of the cassette's "eyes" with a pencil. With great anticipation, I again "play." I recall, with some trepidation, the plot of the *Evil Dead* sequence of horror films. A group of college kids rent an old cabin for the weekend. Inside, they happen upon an old tape recorder with some recorded tapes. They hit play...

9:45 AM

The dial tone fools me into thinking that I am on the phone. Somebody else is dialing. A busy signal, one of those fast ones that seems to indicate that you have made a dialing error. I try to hang up. Can't do it.

9:45 AM

Mostly, I hear breathing. My own.[9] On the fourth ring, a woman, past sixty, answers the phone.

"Hello?"

"Yes, hello, this is Richard Doyle, I am calling from Penn State University to speak with Robert Ettinger. We have been in e-mail contact."

Some breathing, probably not my own. How can silence render perplexity with such precision? "What do you want?"

"We had discussed doing a phone interview."

"Honey! It's a fella from a university."

I told him my name again, told him about the e-mail discussion we had had. "I am doing an essay for a volume on nothing for University of Minnesota Press and I wanted to use some snippets from an interview."

"You were going to call at noon?"

"Yes, oh, I'm sorry, I forgot that Detroit. You want me to call back in an hour?"

"Could you make it two hours?"

"Yes, fine, I'll phone you at one PM your time."

"No, that would be noon my time. Mountain time."

"What time zone is Detroit *in?*"

"I'm in Arizona."[10]

January 1, 1999

I try the DBB, Dynamic Bass Boost, as I listen to myself redial.

There's only so much of this stuff I can review without going into a bit of a panic. What was I thinking? To cite the interview itself, what did I want? Everything—a stray cough, the sound of me taking a sip of ginseng tea—seems to be a sign, including my mistake. Will Ettinger take it as one?

It was an understandable error. Unlike my entirely unwarranted assumption about Ettinger's mortality, my premise that he lived in Michigan was based on some solid textual evidence. The Cryonics Institute, as I mentioned earlier, is located in Clinton Township, Michigan, in interesting but coincidental proximity to Dr. Jack Kevorkian.[11] In addition to being a well-established cryonics provider, the Institute's location gives it a singular status. While some earlier cryonics providers were located in Ohio and Indianapolis, most cryonics organizations today are located west of the Mississippi.[12] I know I would feel better if my freezing—an event whose contingency cannot be erased, bound as it is to the arrival of legal death—took place at a facility close to my death. It only makes sense that distance would be a compelling factor in the choice of a cryonics facility.

So despite all the declarations of the arrival of a "placeless society" that would accompany the virtualization of labor and value, geography would seem to matter—the West Coast ecology cultivates more cryonics than elsewhere. Hence the interest in Ettinger—his Institute, founded in 1976, all concrete, glass, and horizontal—grew at a distance from the informational practices of Silicon Valley, within wrench-throwing distance of that icon of manufacturing, Detroit.

"I'm taping you right now, I'm taping us. Is that ok?"

"No problem."

I was thinking, of course, of Linda Tripp, of what becomes of information when it is grafted into another context, of its essential capacity to, as described by Socrates—okay, Plato scripting Socrates—get into the wrong hands. Clearly, this was not something Ettinger worried about. It was indeed probably something he *couldn't* worry about. Or perhaps it had become something other than a worry.

I wanted to immediately ask him why he was living in Arizona, but I knew it was probably for the weather. And with the global retrieval operations of the Institute, it was good enough to be in hailing distance of the airport. CryoCare, of Rancho Cucamonga, California, was practically located on the grounds of the Ontario airport, and, Lazarus knows, only fifteen minutes from Disneyland.

So I didn't ask him. I also didn't ask him if he had any children, what his sign was, if he believed in a monotheistic god. And I didn't ask him anything about masochism. Yet.

January 2, 1999

10:40 AM

My disk drive has acquired a strange new hum. I drop some Norton Disk Doctor on it. It seems to help.

The sound of a phone being picked up seems only hearable if it is recorded, when you are not listening for it. One seldom hears it from the other side, the tremendous surprise it installs.

I am even surprised the second time.

"Can you make it two hours?"

"I'm in Arizona."

It rings, this time twice.

"Sorry for calling earlier."

"Oh, that's okay."

"I just have a number of questions here that you can either be interested in or not, and we can go from there. How does that sound?"

"Okay."

In addition to *Prospect of Immortality*, Ettinger has written a kind of manifesto for life extension, *Man into Superman*. This 1972 text argued for the research into and deployment of technologies for the eradication of aging. More crucially, Ettinger argued for the use of technology as something other than a remedy for human ills; he saw technology as a vector for human evolution, something he dubbed the "transhuman." In the case of that transhuman practice known as cryonics, I loved the meditative effect of knowing that in order to evolve, one must rigorously do nothing for centuries.

"I wanted to start with your notion of the transhuman. As you may or may not know, there is an emerging school of scholars in literature and the humanities devoted to fostering something other than the human."

Inaudible remark.

"I was wondering if you could touch, talk a little bit about what you see the transhuman as, or what the project is to become transhuman."

"Well, I don't have any particular, specific objectives. I've mentioned a lot of possibilities, in *Man into Superman*, of course. It's just a matter of open-ended progress and individual choice along the way. Well, I suppose there would have to be some kind of amalgam of community choice and individual choice, but in any event there will be choices made along the way, and how it will evolve remains to be seen. All we can say is, in the most general sense, that we want to improve ourselves physically and mentally. As for the course that these improvements take, we'll have to find out."

Amalgam. It sounds like the name of another word, like *anagram*. As I listen to it—the Magnavox/Phillips AQ6581 doesn't have rewind, so I keep flipping it over, fast-forwarding, and flipping—I see a kind of melted mass, as in the ending of *Tetsuo*, when, in a revolution of pleasure, two men are grafted into a machine.[13]

The *OED* tells me, in its tiny print and authoritative way, that I am right to think of a vague, globular mass that refuses reference and connects one immediately to the tangential and the aleatory. *Amalgam* comes from thirteenth-century alchemy, originally perverse:

"Usually taken as a perversion of *L malagma* (in Pliny and the physicians), a mollifying poultice or plaster...A soft mass formed by chemical manipulation."[14]

Francis Bacon later extended this use to include "an intimate (plastic) mixture or compound of any two or more substances."[15] A brain, perhaps? Something flexible—"plastic," and capable of variation. And structurally multiple—it takes at least two to amalgamate.

"That's exactly what I like about your formulation. As you put it, it really does remain to be seen. It's an ongoing, variable process—you don't take a target and then try to orient yourself toward it. It's more a matter of perhaps being available for variation." I was swiping some of this from my skewed reading of Lynn Margulis's work on symbiosis.

"Well, we certainly know something about what we want to *lose*."

Laughter. "Or shed."

"In the form of handicaps, or built-in disadvantages. Obviously, we'd all like to improve our memories, our speed of thinking. We want to lose our vulnerability to disease as far as the physical side of the body is concerned."

Coughing, then a clearing of the throat, as if he were trying to get someone's attention.

"Excuse me. And of course we want to lose our tendencies to irrationality and neurosis and psychosis and skewed emotions and that sort of thing, and we want to lose, oh for example, some of the physiological reactions in fear, we want to change fear from something that is essentially counterproductive to something that, effectively, is just a signal. I mean it can also have bad physiological effects; people can die of fear."

"Stress might be another name for that."

"Yeah, right, stress."

"I am wondering . . . so you think that mostly this is about shedding or losing attributes of our bodies that are negative? Would that be right?"

"Of course the mental and the physical are intermingled. There's no clear line of separation."

Yet another amalgam. "So you would think quite differently than uploaders, for example?"[16]

"I certainly do look forward to positive changes, but as far as uploading is concerned, I don't think that that has even been shown to be a possibility. I mean there's a possibility in the sense in that it has not been ruled out, but I don't think it's a possibility in that we know that an electronic system could be aware or that replicating a person in an electronic system would constitute survival."

"But we do know that for corporeal instantiations of such a system?"

"We do know what?"

"But that we do know the criteria of survival for a corporeal system, like cryonics?"

"No, we don't even know that, we don't know what the correct criteria of survival might be, or even if there is such a thing. Now obviously we have to proceed on the assumption that we survive on a day-to-day basis in the ordinary course of events. That may not be true either. But it's hard to see any alternative to assuming that it's true, so we have to assume that."

"That remains to be seen. But it seems like the presence of the body is persuasive to you in a way that an electronic system is not, for no other reason than that we work *now*."

"The map is not the territory, a copy is not necessarily the same thing, we don't know if isomorphism is enough."

3:30 PM

In *Evil Dead*, of course, the tape turns out to be some kind of incantation, decrypted, appropriately enough, from Phoenician tomb inscriptions. As the tape passes over capstan and playback head and becomes signal, the undead are summoned. The transduction of the tinny speaker commands them into life. Only the complete dismemberment of a body possessed by the undead arrests their force—information theoretic death. The tape may not be the territory—it is not quite itself the undead—but it certainly has its effects.

In the case of my Sony HF 60 type 1, things were not so simple. What sort of interview—networked with my fingers, my Powerbook 520, my continually distracted self—was it summoning?

"You are talking about an attempt to evolve, would you agree with that? If there's an ethic, it's that any opportunity to evolve must be seized."

I fight off the temptation to go back to the *OED*, our codebook, for "seize." All I know is that when a motorcycle, for example, somehow loses its oil, the pistons expand from the friction and heat, swelling until they can no longer move. My friend Bob Felmey had seized his Yamaha, skidding suddenly and getting tossed over the bars. I won't even get into what would be involved with an evolutionary seizure.

"Well, yes, we have to look at our options. The goal is to be somebody improved over what we are now."

"Well, one of the ways we might become is in fact to overcome individuality. Not to drag Nietzsche into it—he is, after all, dead—but Nietzsche saw individuality as a kind of symptom. In this context, to become transhuman would involve inventing new forms of relation, new forms of thought that are something besides 'autonomous' or 'individual.' What remains to be seen is how good the individual form is for inventing or surviving the transhuman."

"Well, as you know, it's a common theme in sci-fi, some people would say that we are already communities, that the individual organism, as we think of it today, is actually a community."

Yes, yes, literature, sci-fi, pop culture as the software of a community based not on identity—who agrees about what they read?—but about the capacity to be affected by others, even if that other be a horror film. Somewhere in the background is typing. I'm pretty sure that I closed the door to my office when I taped the interview, but it could be coming from a colleague's office or it might be in the background. A typewriter in Ettinger's home. What is it typing? More sci-fi?

"Are you thinking of people like Minsky, *Society of Mind?*[17] What do you think of that idea?"

Ettinger reminds me that he has written about something called the "self circuit," what he takes to be the ground of being. Such a conception is closer to the Ouroborous—a snake swallowing its tail, a homeostatic mechanism—than to a swarm.

"So we have something like this, as a vague suggestion, that it is a standing wave—*coughing*—responding in various ways to stimuli of various kinds." *A throat clearing.* "If that's true, whether that could be duplicated in some sort of community mind or hive mind—*cough*—that remains to be seen." *Coughing.* "I'm sorry, I've got to get a cough drop."

Coughing. That noisy phone silence, where any trace of sound signals a return to the phone.

Has he forgotten me? I feel regret, as if I had ignored the physical toll it extracts to talk and talk and talk.

Sixty-five seconds have gone by. It sounds like he is about to talk. I respond to the silence. I clear my throat. There is more coughing.

At ninety seconds, "Hello?" Silence. "Oh . . ." False alarm.

"I am sorry to keep you waiting."

Surprise. "Oh, no, no problem."

"Yes, I was about to say that many philosophers and investigators tend to assume that it is not only desirable but possible to find the answers to what they are looking for on the basis of what they already know. Unfortunately that's not the way the world works. We have to enlarge our database by experiment before we are even in a position to achieve some of those goals of understanding, particularly in biology. The question of criteria of survival and so on is not a philosophical question, at least not entirely, and not in the sense that the older philosophers thought. It's both a philosophical and a biological question."

Again, I am thinking of the compound mass, amalgam. I like the way Ettinger respects the sheer difference of the future. In a way, cryonics—that rigorous operation of nothing—involves a disciplined willingness to be surprised, an involvement with the future based on what is not only unknown, but, in the present, unknowable. Bundled with, compounded by, the future, cryonics involves continual and uncertain waiting.

January 4, 1998

9:30 AM

By now the cold had intensified. Those dusty flakes had become gobs of sleet. An icy rubble littered the roadside as I foolishly pedaled my bike toward campus, frostbite creeping up on my digits. I have a new theory about why cryonics is more popular in California.

The batteries are starting to run down on the Phillips/Magnavox AQ6581. I confess that I have been listening to something besides Ettinger. William S. Burroughs collaborated with Material to produce a spoken word-cum-musical amalgamation based mostly on *The Western Lands*, his last novel. Burroughs's late work is intensely concerned with immortality, but Burroughs located a recipe for immortality distant from the usual understanding of an eternal repetition of the same—Ettinger's "it would be enough simply to continue to live for a few centuries the same kind of life I'm living now, that would be just great." Instead of a sturdy continuity, Burroughs sought to practice an immortality for "soft machines"—a radical increase in the capacity for mutation:

> Immortality is prolonged future and the future of any artifact lies in the direction of increased flexibility, capacity for change and ultimately mutation. Immortality may be seen as a by-product of function: "to *shine* in use." Mutation involves changes that are literally unimaginable from the perspective of the future mutant.... Mutation is not a matter of logical choices.[18]

Ettinger's voice was still there—fainter, a half-octave lower maybe, but there. We talked about the possibility—following up on the idea of individuals as communities, swarms—that Ettinger's "self circuit" could be distributed—centerless and made of connection rather than location. Ettinger said it remained to be seen.

"Well, it raises the question of what you would freeze. I mean it could be that this basis of identity is not even located, strictly speaking, in the body. It could be an ecological event."

"I'm not sure I follow you."

"Well," now *I* am coughing, "most organisms we consider as emerging out of an ecology. They evolve not as individuals but in relation to other organisms. Since I assume that the self-circuit evolved, then the question arises whether or not this 'self' can be localized in an individual human body or if it is an event that emerges in an ecology."

Ettinger wasn't having any of it. "Well, no, I certainly cannot see any way that one could arrive at the conclusion that I don't exist in isolation."

Silently, I try to think of a way to bring Ettinger to exactly that conclusion. It turns out that this silence might just have done the trick.

"Now, of course, there are hints in physics that *vaguely* suggest that people may be connected, I'm sure you have seen a lot of these recent reports that indicate that there are so-called quantum entanglements between any two systems that have ever interacted, and some people interpret that to mean that any two systems no matter how widely separated, if they ever interacted in the past, they are still connected physically, through quantum entanglements. That leads to some very vague and very distant speculation that people also could be entangled with each other, somehow, but it is extremely vague so far."[19]

"Yes, emphasis on the 'somehow,' probably."

Laughter. More sound of typing.

I take the tape out of the AQ6581, searching it for signs of entanglement. The fold had not returned. "Somehow," Ettinger was implying, we were now, possibly, entangled. I would just have to wait and see. I had forgotten to ask about the "archive" question—cryonicists store tapes and CD-ROMs along with their bodies to "remind" them who they are upon revival. I would send a copy to him for the freezer.

2:30 PM

How long does it take for a man to realize that he does not, cannot, want what he wants?

William S. Burroughs, *The Western Lands*

I ask Ettinger what he thinks of science fiction, what its role was in the ongoing practice of building cryonics. I knew from the introduction to *The Prospect of Immortality* that he had published his own science fiction, ("The Penultimate Trump," *Startling Stories*, March 1948) and that science fiction pioneer Hugo Gernsback had influenced him as a child. After trading examples of cryonics in literature—I forgot Frederik Pohl, Ettinger left out Philip K. Dick—he said something that surprised me. Ettinger claimed that science fiction actually had a negative effect on cryonic practice. He confessed that his own inquiries were spurred by a response to science fiction, but he worried that cryonics' fictional representation often convinced people it was indeed fantasy. It made them forget that cryonics is possible. Since I think of cryonics as, at least in part, a fan practice of science fiction, I listened and relistened to this one several times. Hard on the batteries.

"We're talking about science fiction as a way of experimenting with cryonics, with making it thinkable, even marketing it."

"I have often had the opposite thought, that, as far as science fiction is concerned, that there has been a negative effect. There's what I call the inoculation effect. When people read about something over a long period of time in the context of fiction, then they assume it *is* fiction."

Laughter.

"People have been reading about it, or seeing movies about it for so long that the net result in their minds is that "this is fiction." The cryonics organizations, not ourselves but Alcor for example, for a long time would send delegations to science fiction conventions, set up booths and try to make sales, and the results were abysmal."

How would Merkle's cryptographic scheme deal with this problem, the "inoculation effect"? If the rigorous future decoding of the scrambled text of the body relied on the premise that, with freezing, such a body was rendered more readable, how would any such protocol deal with the problem of fabulation that Ettinger was gesturing toward here, the tendency of texts to continually mean something else? Whereas science fiction texts aren't, on Ettinger's terms, "referring" to cryonics at all—they are creating possible worlds in which cryonics operates, not representing cryonics in the present—they nonetheless seem to do so continually: they refer to cryonics as fictional, they rigorously produce a sense of cryonics' falsehood. Given this, after "a long time"—for by any account, cryonics must remain science fiction for a long time—who would be able to take a cryonic corpse seriously?

Maybe there will indeed be work for English professors.

3:45 PM

The batteries are really low now. I try to conserve them by rewinding by hand, with a pencil through the left eye of the cassette. As I do so, I can see the tape loosen inside, form a fold, then snap back taut. I don't like the stress on the tape, but I've got to finish. Today.

I ask him if he thinks our concepts of death are less mutable than our understandings and practices of sexuality. I'm still trying to muster the nerve to ask him about cryonic sexuality. More about that later.

"We have seen great change in the way in which we view sexuality but relatively little change in the way we view death. I suppose there's a good reason for that. Do you think that there is something about death in American culture that is particularly intractable?"

"Well, we have lived with death for thousands of generations, so the survivors are those who have at least the capacity for coming to terms with death, which is another way of saying that people have these built in mechanisms, either individual or social, communal, that allow or possibly even compel them to come to terms with death."

I am breathing really loudly now. The rhetorician in me loves "living with death," as if it were some annoying if incessantly absent neighbor that, in the end, one just had to accommodate oneself to. I realize that anticipation is the practice of the moment—what else could "living with death" entail besides being capable in the future of mourning?—so I try to introduce my question about masochism.

"I'm trying to think about the effect of preparing to be frozen, the effect of cryonics on someone who is thinking about being frozen. I am thinking about it in the context of some of your discussions of sexuality in *Man into Superman*, where you foresee 'A perpetual grapple, no holes barred . . . continuous state of multiple orgasm. (Of this, more later.)'[20]

"What I think is so remarkable about that is the way in which *deferral* itself—'of this, more later' is an important component of describing the "perpetual" or immortal character of this sexuality. Indeed, in some sense it *is* a sexuality, this practice of deferral. I wonder if what it means to be a cryonicist is to experience . . . to do deferral differently than others. We all anticipate our deaths in some ways—"

"Well, possibly, let me say at least one thing before I forget it. I think there's going to be a substantial change in the relatively near future when people begin to take seriously the thought that their grandchildren, maybe even their children aren't going to die of old age. It's one thing to think of yourself as just another chain of generations, and that that's part of the eternal scheme, but if it starts to dawn on you that you might be one of the last generations of dinosaurs, that's a little bit different."

"What do you think that would be like, to be one of the last generations of dinosaurs?"

"Well, I don't think people would be happy to know that they're going to die, but their children and grandchildren won't."

"They'll be jealous?"

"They'll be jealous and they won't think of themselves anymore as part of an infinite succession of generations, and they won't think of themselves as honored in their turn, as they honor previous generations, to the extent that they do."

"It would be a kind of complete forgetting."

"Yes, it will dawn on them, that, essentially..."

January 5, 1999

What characterizes masochism and its theatricality is a peculiar form of cruelty... the

specific freezing point, the point of idealism realized.

Gilles Deleuze, *Coldness and Cruelty*

Ice, it turns out, creates linkages, even hospitality. I'm walking from the parking deck to my office. I need to finish up a few things, print it out. The AQ6581 still chugs along, and our slightly slurred voices divert me from the cold of the walk.

"So can you respond to the notion of deferral? It seemed like in your discussion of sexuality, that what seemed pleasant about that sexuality was that it could be deferred, it would never be over in some sense."

I certainly wasn't being clear. Not everyone connects the use of technologies to an understanding of desire. I was try to figure out how to decode this concept, to connect it to something Ettinger already does.

"I'm afraid that I am getting this tickle back in my throat."

"Oh, I'm sorry. Would a glass of water help?"

"I'm afraid we're going to have to stop for today."

"Well, I hope you are okay."

"I'm fine. Sometimes it happens when I talk for a while."

"Thank you very much, I hope to talk with you again soon."

"Yes, we can talk another time. You're welcome."

"Bye-bye."

Over the past few months, the landscape of Central Pennsylvania has been littered with figures whose presence in this narrative would effectively

dramatize my encounter with Robert Ettinger. There was even an ice sculpture con-
test on New Year's Eve, part of an anti-alcohol, family-entertainment New Year celebration sponsored by my town. Templates were drawn on huge, blank sheets of
paper, detailed schematics grafted to slabs of ice through the very adhesion of the
cold. A group of men with chain saws carved the lines with their screaming, frenzied
blades, while others watched as vague, glistening forms emerged, dripping. But
that's another horror film.

      All of the other ice—forged from a slushy muck of salt and anti-
freeze that was finally arrested by the deepening, increasingly painful cold—preserves a more diverse array of singular moments, events not usually linked. A pack of
matches, its cover unreadable, flaps out from a berm of chunked ice. There are still
oak leaves, flaking from the cold, scattered across the expanses of asphalt scoured
from the rubble. Headphones on, listening to the new sound of Ettinger's absence, I
slip on a patch of ice, two steps without traction, doing my best Chaplin, and I am
down.

      And the cold is here now, for a long time. This amalgam fuses an
ecology whose contingent order remains to be seen.

S   I   X

Uploading Anticipation,
Becoming Silicon

...more life, fucker!

Blade Runner

WHY RESPOND with such an outrageous demand, an expletive that, if Nixon had said it within his relentlessly taped Oval Office, would be deleted from any transcript? Why this demand from a machine?

It is not so clear these days what a machine *is*, exactly. The transformations of the life sciences and its cultural ecology that I have attempted to narrate thus far have turned the organism/machine opposition into a smudge, the topology of which I have described as "postvital." No longer a sovereign site of interiority, a vital inside that struggles with an inanimate outside, organisms in the contemporary life sciences become moments in an evolutionary syntax, nodes in an network not of "vitality" but of information. More than a becoming-machine of the organism, this retooling or "refiguring" of life provokes double takes on the becoming-lively of the machine. "Our machines are disturbingly lively, we ourselves frighteningly inert."[1]

Thus the demand from the machine—"more life, fucker"—becomes not simply an index of the desires machines have today, but a map of the odd rhetorical imbroglios of "life" that traverse both contemporary science and science fiction. A demand from a machine, "more life, fucker" maps the transversal movement of that long infection, "life." It is this ability of the machine to be the double of life that "more life, fucker!" articulates, a capacity to simulate that enables such remarkable production of knowledge about living systems.

This informational nexus between organisms and machines could be described in two registers. On what we might characterize as the rhetorical plane of technoscience, living systems themselves "double" or simulate the flow of information that enabled the nearly unthinkable expansion of knowledge associated with molecular biology. Thus the conceptualizations of "life" associated with the meta-phorics of "code" help craft a set of knowledges to the extent that *organisms* can simulate *machines*. The debt this crafting of knowledges owes to the rhetorical practices of molecular biology is not easy to evaluate, but one of the rhetorical effects fostered by the notion of "code," for example, was mobility: rather than tied to the interior of a sovereign organism and its life, the essence of an organism was, like any other code, able to be deployed in any context, deterritorialized.

It is this mobility—the ability of living characteristics to move from one ecology to another through other than organismic routes—that instantiates the other register of relations between organisms and machines. If molecular biology "folded" organisms, rendering them for their codes, moving backward before the "unfolding" of development, from flesh to its code, this mobile rendering of the organism also launches a *redistribution* of vitality, a simulation of life by machines. The very rhetoric of code that transformed the life sciences allowed the code of life to find an ecology in silicon, artificial life.

This deterritorialization of life—its connection to milieus other than carbon—is neither a simple reification nor a reduction. Feminist psychologist Elizabeth A. Wilson reminds us that a distributed or connectionist notion of cognition—one in which cognition emerges not from a location but *between* locations, a network in which "each unit gains its identity not through any essential characteristics...but through its placement in the connectionist architecture"[2]—offers enormous promise for understanding ecologies of transformation or learning, informatic economies that are irreducible to a set of rules or algorithms. This connectionist map of cognition renders a human subject sculpted not from presence but from difference, an incessant variation whose contingency is directly related to "connectedness"—its exposure and relation to something other than itself. So too does a distributed model of life—in which organisms are effects of acentered networks rather than privileged locations of vitality—foster the encounter with life which is something other than an "essence" or a sacred site of interiority.

But why the curse associated with the investment of machines with vitality, "more life, fucker"? If we remember, or deploy some false memory implants of the sort referenced in this quote's origin, this demand is directed by an android to its wetware creator, Tyrell, the ambiguous "father" of the machine. Such a memory replays the truth of simulacra: they replicate rather than reproduce, proliferating through an assemblage of repetition and materiality—an entire ecology mobilized for iteration. Replication is the simulacrum's habit and habitat. "Fucker" is therefore not a term of abuse, but an empirical observation about the differing modes of proliferation deployed by simulacra and sexually reproducing organisms respectively; simulacra are not heterosexual kin, they threaten to float free of their economies or ecologies of "origin." As replicators and not reproducers, the "skin jobs" of *Blade Runner* threaten not to fuck. Simulacra emerge out of an ecology of neither sex nor death, and the subsequent representations of dangerous desire—a desire not anchored in any reproductive economy—is tellingly indexed by the logic of Rachel Ward's response to a test to determine if she is an android: "Mr. Decker, is this a test to determine if I am an *android* or a *Lesbian?*"

This threat—simulation's ability to overtake or "forge" the original and become untethered—provokes the demand for "more life." *Blade Runner* anticipates the contingency of a machine that would float free of the constraints suffered and enjoyed by its creators. As such, androids in the world of *Blade Runner* have built-in death "effects," an off switch that automatically engages after four years, attracting simulacra to the same tired trope that incites *Blade Runner*'s fleshier characters—death. *Blade Runner* is thus less a narrative about the latent vitality of machines and their proliferating simulacra and more a testimony to the virulence of that human virus—lack. For in its formulation of "more" life, the android replicates a discourse in which life is continually found wanting, one in which the future is encountered as a threatened and entropic depletion of the present.

This evacuation of the present by the future often figures in tales of exhaustion and forgetting: "We have forgotten the organism. We have forgotten the body." Perhaps these mantras bear repeating; the contingent emergence of molecular and computational thinking is hardly a simple occasion for celebration. Yet if the implosion of organisms and molecules, computers and bodies, dislocates and distributes our corporeal and cognitive effects, it does so through an encounter with the unprecedented, an exteriority whose arrival resounds with more than loss. Through the veritable deconstruction of the concept of life by molecular biology we have swapped an essence for becoming; vitality emerges between the nodes of a network, intermezzos oriented toward the future—"It's evolving"—but constituted out of contingency—"what is it becoming?" Less a loss of corporeality than its deterritorial-ization, the becoming-informatic of life reminds us of the enormous capacities for difference that are living systems. Each new pragmatics mobilized by a molecular biology founded on the "secret of life" foregrounds the fact that, to paraphrase Gilles Deleuze sampling Spinoza, we don't know what life can do. In short, new configura-tions of life are new organizations of contingency, other substrates for becoming.

These new morphologies of becoming provoke Deleuze's inquiry into the Nietzschean figure of the *übermensch*. Neither "god" nor "man," the *über-mensch* paradoxically emerges out of new relations to finitude. Deleuze maps this composition of human morphology in topological terms, describing the historical emergence of infinity in terms of its capacity to "fold" man, to give humans an "in-side," an infinite soul. Deleuze distinguishes this topological form called "man" from that which emerges from the "superfold":

> It would no longer involve raising to infinity or finitude but an unlimited finity, thereby evoking every situation of force in which a finite number of components yields a practically unlimited diversity of combinations. It would be neither the fold nor the unfold that would constitute the active mecha-nism, but something like the *Superfold*, as borne out by the foldings proper to the chains of the genetic code, and the potential of silicon in third-generation machines, as well as by the contours of a sentence in modern literature, when literature "merely turns back on itself in an endless reflexivity."[3]

Deleuze describes the emergence of the *übermensch* not through Foucault's visual ana-lytics of the "disappearance" of man but through a topological turning or troping of man. Crucial to this superfold is its pragmatics—only as a practice does its "unlimited diversity" come into play. Hence the practices proper to the superfold entail an irre-ducible contingency—the contingency of enactment. As with Wilson's analysis of a distributed cognition, the distributed vitality of the superfold—"the potential of sili-con"—is irreducible to any set of rules or algorithms, as it "resides" only on an itin-erary—becoming.

If such contingency organizes the narrative of a potent site of science fiction—within *Blade Runner*, the inability to distinguish human from simu-lacrum is unprecedented, the future itself—it also inhabits contemporary narratives of technoscience that themselves blur the border between science and fiction—sci-ence/fiction. Uploading, the future porting of human identity and corporeality to a noncarbon substrate, is a contemporary utopian narrative of becoming-silicon, a set of rhetorical operations that render the future as "more life." Copied onto a higher level instantiation, uploads herald the arrival of a technobody built with the future, an exfoliation of the brain onto the universe itself. Indeed, physicist Frank Tipler, in *The Physics of Immortality*, argues that uploads will eventually *engulf* the universe.[4] Such utopian narratives, announcing a future of plentitude and immortality, figure the implosion of flesh and silicon as an excess, a transformation of the human ecology that promises to materialize infinity, to kill death. What technologies, rhetorical and otherwise, enable these narratives? How does uploading broker the new alignments of information, bodies, and subjectivity and render not loss but transformation? Aren't these fuckers simply asking for more lack?

Hacking Lack

We might be tempted to locate an amnesia of the body in any conjunction of computers and human subjectivity. We might emphasize the way in which the discovery of DNA as a site of memory has, ironically, fostered a forgetting of corporeality.

Critic Scott Bukatman, in his remarkable book *Terminal Identity*, narrates precisely such a loss of corporeality in the discourses of science fiction and cultural theory. As storytellers of the spectacle, Bukatman notes, cyberpunk scribes allegorize the disappearance, dissolution, or implosion of the body in technoscientific economies of simulation and informatics. In such regimes, "The body exists only as a rhetorical figure."[5] This collapse into rhetoricity, where bodies are rendered as occasions for coding, is marked by Bukatman as a loss, a deficit incurred by becoming postmodern. This lack extends not merely to the increasing devaluation of the experience of the body—as in the notion of genetic "disease" or the allegedly disembodied state of cyberspace—it troubles the very ontologies networked with the corporeal habit of subjectivity. Describing the unmoored character of both postmodern science fiction and "theory"—the "narratives of terminal flesh"—Bukatman locates yet another lack:

> The narratives of terminal flesh offer a series of provisional conclusions wherein the subject is defined, at different times, as its *body*, its *mind*, or sometimes its *memory*. This proliferation of definitions reveals the absence of definition: our ontology is adrift.[6]

While Bukatman's analysis could be coupled with my own claim that the contemporary life sciences have smeared the operations of organisms and machines, his argument also underscores the sense of loss installed by the mobility of bodies and codes. But such an analysis—with its revelations of "absence"—functions as more than nostalgia for the allegedly stable ontologies of the past; it forestalls the affirmations of multiplicity—drifts—that are *also* at play in late, late, late capitalist science fiction and theory. The transformation of a "proliferation" into an "absence" is more than a theoretical error, an inability to narrate the enormous transformations in subjectivity and corporeality that Bukatman so admirably relates. Such nostalgia also forestalls the effects that inhere in the very transformations wrought by the technoscientific renderings of the body, contingencies that are not merely represented in science fiction and theory but *provoked* by them. For clearly, these "narratives of terminal flesh" cannot be simply extracted from the new configurations of bodies and machines that they render. As Allucquere Rosanne Stone has argued, for example, William Gibson's *Neuromancer*—with its terminal flesh trope of "meat"—became the conceptual currency that supported virtual technology research in the early 1980s.

> The critical importance of Gibson's book was partly due to the way that it triggered a conceptual revolution among the scattered workers who had been doing virtual reality research for years...the technological and social imaginary it articulated enabled the researchers in virtual reality—or, under the new dispensation, cyberspace—to recognize and organize themselves as a community.[7]

Science fiction and theory are thus not merely narrating or "representing" transformations of subjectivity and corporeality; as rhetorical softwares they constitute, discipline, and organize scientific communities with such retoolings of the body and the self.

Uploading, taken as a practice and not simply an imaginary ideal, a symptom of ontological loss, becomes what Foucault has characterized as a technology of the self, a comportment or "fashioning" of subjectivity humans carry out through "a certain number of operations on their own bodies and souls, thoughts, conduct, and way of being, so as to transform themselves in order to attain a certain state of happiness, purity, wisdom, perfection, or immortality."[8] These "technologies of the self" are primarily discursive, as in the confession, where the obligation to speak the truth about oneself, and to exteriorize it in the act of confession, paradoxically constituted the interiority of Christian subjectivity. Foucault linked this confessional discourse with the growing importance of writing to the emerging bureaucracy of Rome:

> Taking care of oneself became linked to constant writing activity. The self is something to write about, a thing or an object (subject) of writing activity.

This is not a modern trait born of the reformation or of romanticism; it is one of the most ancient Western traditions. It was well established and deeply rooted when Augustine started his *Confessions*.[9]

Under this rubric, the secret interior of Christian subjectivity—the hidden failure that must be exposed—is constituted not out of repression but through the productive rhetorical circuit of confession. In this economy, the subjectivity of the confessor becomes a mere node in the discourse network of Christianity, her flesh the materiality of the writing substrate through which this new soul emerged, was delivered.

But this economy was not of just any character; the discourse network ran on the tropes of finance. While Seneca envisioned self-examination through the lens of an audit, "when a comptroller looks at the books," Cassian writes through the medium of currency. "Conscience," the internal simulacrum of the externalized confessor, becomes "the money changer of the self." Ruthless in its discrimination and sense of authenticity,

> It must examine coins, their effigy, their metal, where they came from. It must weigh them to see if they have been ill-used. As there is the image of the emperor on money, so must the image of god be on our thoughts. We must verify the quality of the thought: This effigy of God, is it real?[10]

Cassian's technology of the self is thus haunted by money's simulacrum, the threat of a copy without origin. Through the continual "verbalizing" of the self, the subject is rendered and purged of scandalous threats to its imprimatur of originality. Master of its own double, the self emerges "in the permanent verbalization of all our thoughts. . . . This verbalization is the touchstone or money of thought."[11] This rhetorical economy—the persuasive character of money, the verbalization of thought—insured the possibility of a noble transaction of selfhood, a transaction with and through the currency of God.

This technology of Christian identity, then, cultivates a transaction space for and of subjectivity, a space exterior to the self that would evaluate and inscribe the self, a subject cultivated through the practices of the market: calculation, evaluation, verification. This "money changer" self operates in a continual present—"is it real?"—and encounters the past only through loss—worn coins, notes of dubious origin.

This continual composition of the self, in its constant encounter with the practices of "value," operates only through an encounter with humans. No alterity—a material, organismic, or technological familiar—enters into the self's habit or habitat. What Foucault characterizes as the "discriminating power" flows through the face and speech of a "master."

> Even if the master, in his role as a discriminating power, doesn't say anything, the fact that the thought has been expressed will have an effect of discrimination.[12]

Hence the discourse that articulates this technology of the self operates as an order, a command which installs a difference—"discrimination"—as its primary effect. This difference is a difference of negation. In the cacophony of the confession or the silence of meditation, discrimination renders a self composed of renunciation. "You cannot disclose without renouncing."[13] The very matrix of this renunciation is the space of evaluation that, in the first instance, determines the contours of the "human"—the tautological ideal of a master in transaction with a subject.

While the fine grain of Foucault's arguments concerning technologies of self and their relations to rhetorical practices calls for more inquiry, I want to map out a crucial characteristic that might help to bring uploading's difference into relief. The technologies of self analyzed by Foucault are essentially based on equilibrium—the confession becomes a discourse franchise on the basis of its regularities; penitence becomes a repeatable, if asymptotic, "model" of martyrdom. By contrast, *Blade Runner*, itself an uploaded version of Philip K. Dick's *Do Androids Dream of Electric Sheep?*, features an anticipated, postapocalyptic world where novel and unpredictable forms of empathy extend primarily to the inhuman.[14] This encounter with the inhuman or ahuman operates by definition as an itineration rather than an iteration, a drifting rather than a marching, what Deleuze and Guattari characterize as a "following" in a discussion of the distinctions between a "royal" and "minor" science, a discussion I take up earlier in this book. In following, Deleuze and Guattari locate a response-ability to a milieu, a capacity to be affected by the finitudes and contours of a given ecology. In this sense, practices of itineration—an always mobile metallurgy that follows the flow of metal and ore, an ambulant computer science that must always, finally, "run" its code—are constituted out of a continual encounter with something other than human, an alterity that cannot be mapped or, by definition, "known" in advance. The structural impossibility of knowing such singularities in advance is hardly a lack—relations with the inhuman take place as an encounter rather than an understanding. It is this possibility of forging technologies of self that are constituted through a regard for an inhuman exteriority that, I will argue, uploading indexes.

More Copies, Fucker: Becoming-Sampled

Unlike the technologies of self described by Foucault, contemporary technologies—
rhetorical and otherwise—provoke something other than an authentic, individuated
self constituted through originality. Writing of the possibility of replicating humans
through cloning, William S. Burroughs characterizes the technologies of identity as
an "illusion," and urges on the possibilities of multiplicity:

> The illusion of separate inviolable identity limits your perceptions and con-
> fines you in time. You live in other people and other people live in you; "vis-
> iting" we call it and of course its ever so much easier with one's Clonies...I
> am amazed at the outcry against this good thing not only from men of the
> cloth but from scientists...the very scientists whose patient research has
> brought cloning within our grasp. The very thought of a clone disturbs
> these learned gentlemen. Like cattle on the verge of stampede they paw the
> ground mooing apprehensively..."Selfness is an essential fact of life. The
> thought of human non-selfness is terrifying." Terrifying to *whom?* Speak for
> yourself you timorous old beastie cowering in your eternal lavatory.[15]

Burroughs has, of course, been among the most energetic terrorists of multiplicity.
But his affirmation of replication in the context of a herd of lack cattle—*more copies,
fucker*—marks the possibility that the multiple ontologies of the present yield something more than a lack of selfhood or humanity. For Burroughs, both the desire for
immortality and the terror of replication emerge out of a bad joke: the self. Less a
foundation than an allergic reaction, the self is for Burroughs a pathetic "swelling"
with no more continuity than "a fever sweat." As in the "Clonies" example above,
Burroughs gestures toward a distributed identity, one that resides as much in others
as in one's "own" body. Less an extinction of the self than its deterritorialization,
Burroughs's self is a becoming. "Terrifying to whom?"

No wonder, then, that Burroughs describes conventional visions
of immortality as wrong. As multiplicities, humans garner their identities through
transformation and adaptation rather than mastery and autonomy. For Burroughs,
the possibilities of cloning are fascinating precisely because they shatter the originality and authenticity of the ego, and foster new forms of individuation scattered
across bodies. The trick in getting out the hell of Time, as Burroughs would have it,
is to become a transversal "body" broadcast across space. Organ transplantation and
cloning thus become not "vampire schemes" for the overcoming of entropy but instead compose tactics for becoming other, the distribution into space.

The very operation of Burroughs's writing attempts to provoke just such multiplicity. The cut up—literally chopping up text and rearranging it—decomposes the consciousness of the "author" Burroughs, even as it distributes the Burroughs effects across texts, films, bodies, and music. The Burroughs effects thrive precisely to the extent that they harbor a capacity for difference—the ability to encounter the difference of other contexts, to be affected and transformed by them. Paradoxically, then, Burroughs's route to immortality is to become other, adrift.

Uploading, as a technology of the self that is also such a technology of replication, operates through the transaction spaces made available by global capital and its emerging writing practices. Deferred to the future, the practice of uploading nonetheless mobilizes its own shattering effects: dislocation through anticipation. In defense of the plausibility of uploading, a poster on CryoNet, a discussion group devoted to cryonics and other immortality technologies, highlights the "virtual" character of money and work in the present:

> I've also never had a "real" job. Work consists of force times distance. I do some work on a keyboard, but it's a negligible amount. My employer could save money by replacing me with a steam engine which would press all the keys with greater force, through a greater distance, more often, thus doing far more work than I ever did. Actually, this wouldn't save them any money at all, as they never pay me money. They only give me pieces of paper that have pictures of dead presidents on them. Come to think of it, they don't even do that. They give me pieces of paper which tell me that numbers in a bank computer somewhere have been incremented. Which is silly, because numbers can't be in a computer, as numbers are just an abstraction. Actually, they merely change the state of magnetization of various very tiny areas on a disk belonging to a bank. Why should I do real work for that?[16]

In the relentless calculations and anticipations of uploading that take place on Usenet groups like sci.nanotech and sci.cryonics, the evaluation of the uploaded self emerges out of an economy not of individual "money changers" who can determine the veracity of the "money of thought" but through the new mobilities of global markets. Both labor—the hilarious vision of machinic replacement, itself substituting for a thousand monkeys on a thousand typewriters—and value—the shifting and fleeting movements of magnetic media—are figured as elements in an economy of transformation, a regime not of the real but of "change." Indeed, in this milieu the "real" has become kitsch, a ridiculous throwback as primitive and charming as a steam engine.

The practices of everyday life, and the technologies of self that emerge through them, are thus composed not so much of evaluation and verification as of anticipation, the sometimes anxiously performative constitution of futures. This ecology of self flows with the "change of state"—machines transform human labor, numbers are incremented, disks become reorganized. More than a displacement of the real by the virtual, this narrative highlights the matrix of constant transformation within which the discourse on uploading is articulated.

Indeed, this association of the alleged abstraction of virtuality with the continual variation of global capitalism reminds us that virtuality is anything but unreal. Virtualities may lack reference, drifting without anchor in the actual, but it is precisely in the contingent relation of virtuality to actuality—becoming—that new capacities to be affected, futures, emerge. Virtualities are less the ethereal precursors to actuality than the positivity of transformation itself, diagrams of futurity. To encounter the virtual is to enter into an itinerary, an intermezzo Deleuze characterizes less as an intentional effort and more as a capacity to be affected, an active drifting or surfing: "The basic thing is how to get taken up in the movement of a big wave, a column of rising air, to 'come between' rather than to be the origin of an effort."[17]

The uploaded subject, as the technology of self that effects the writing of subjectivity not to flesh but to silicon, thus marks a relation to market practices based not in the present or the past but the future, the continually itinerant specter of a "change in state." No longer referenced by the transcendental guarantee of gold, the "conscience" of global capital emerges not from authenticity or originality, but through the flickering of money's double: the future. As mathematician and semiotician Brian Rotman has argued, the "scandal" of the post–Bretton Woods dollar is that it is "backed" only by its ability to be traded on a secondary market in the form of futures contracts. This "xenomoney" is

> floating and incovertible to anything outside itself... Its "value" is the relation between what it *was* worth, as an index number in relation to some fixed and arbitrary past state taken as an origin, and what the market judges it *will* be worth at different points in the future.[18]

Thus the value of money, rather than keyed to its authenticity or originality, is constituted by the future. It is the *anticipated* value—"what the market judges it *will* be worth"—that drives the operations of futures markets, Cassian's "money changers" of the present. Rotman emphasizes the "loss of anteriority" that inheres in such a shift in the status of monetary signs, but, as with my discussions of cryonics, I seek to map the ways in which such markets operate through the rhetorical production of anticipation, an active drift of anticipation that orients uploading's technology of self toward the future, a future of itself.

To be sure, "uploading" exists as an anticipatory technology of the self. Like Burroughs's "Clonies," the technologies of silicon brain or body replication rely on material practices of the future. Robot surgeons, nanotechnological assemblers, and extraordinarily high-resolution electron microscopes are among the gadgets that can be deployed only contingently, in a possible future. But technologies of the self are not, of course, simply *technical*. As in confession, where the discourse network of Christian subjectivity operated through the disciplinary matrix of the "money changer," uploading operates in the present through the rhetorical production or "uploading" of anticipation, a summoning of technologies from the future that would constitute a self available for replication and immortality. While some forms of Christian subjectivity, according to Foucault, produced interiority through the masterful exterior relay of the priest, uploading's technology of self forms a discourse circuit with the future, the arrival of differences that can only be followed, encountered as an itinerary rather than an iteration. Though the nanotechnology that would enable the uploading of human identity resides in a possible future, the subjectivity formation that would produce and inhabit such technologies sprouts in the itinerant, anticipatory ecology of the present. Indeed, in some sense uploading emerges as the fantasmatic extrusion of a twenty-four-hour global market, a market that demands a subject whose very body is money.

On the Other Thumb...

In William Gibson's *Neuromancer*—that apparent obligatory passage point for any discussion of "cyberspace"—such an anticipatory immortal replicant is figured as Dixie, the "flatline construct," a computer replication that renders the skills and effects of a hacker long dead.[19] While Case, the appropriately named container for drugs and hacking desire, has his very body leveraged by corporate capital—he must complete the job to avoid the dissemination of toxins throughout his body, toxins that will forever render him unfit for the ecology of cyberspace—Dixie's leverage is otherwise. In exchange for networking with Case and the completion of the job, Dixie demands the right to be erased: "This scam of yours, when it's over, you erase this goddamn thing."[20] Dixie, in the grammatical third person—"this goddamn thing"—anticipates a future of absence. *No life, fucker.*

What's crucial to note in Gibson's figuration of an upload—a figuration that helps render uploading thinkable within the community of Transhumanists, Extropians, and other aficionados of uploading—is the role that Dixie's "body" plays in the transactions of *Neuromancer.* In this economy, even uploads get paid. In Dixie's second death, the currency of the transaction is nothing other than his phantom body—its absence is the price that Case and Dixie's employers pay for work completed. Thus for Dixie it is possible not only to examine himself *as if* he were currency; he *is* currency, the currency of a transaction that would allow his exit from the money economy, i.e., out of any economy whatsoever, this "scam."

Dixie's currency, though, suffers the uncanny fate of the simulacrum. Like Cassian's technologist of self, that money changer called "conscience," Dixie ceaselessly evaluates the authenticity or "reality" of his own thought:

> "How you doing, Dixie?"
> "I'm dead, Case. Got enough time in on this Hosaka to figure that one."
> "How's it feel?"
> "It doesn't."
> "Bother you?"
> "What bothers me is, nothin' does."[21]

But rather than merely experiencing this "nothin'" as a lack, Dixie notes the uncanny, itchy presence of the simulacrum, comparing his fate to a phantom limb:

> "Had me this buddy in the Russian camp, Siberia, his thumb was frostbit. Medics came by and they cut it off. Month later, he's tossin' all night. Elroy, I said, what's eatin' you? Goddamn thumb's itchin', he says. So I told him, scratch it. McCoy, he says, its the *other* goddamn thumb."[22]

An itch that can't be scratched, the uploaded subject can only anticipate. Rather than an incessant "verbalization of thought," though, Dixie's technology of self operates through the anticipation of something other than the simulacrum, the positive production of erasure, a fracture in the endless iterations of selfhood. In his desire to erase his presence, Dixie seeks to comport a subject in the future that would escape the nonstop operations of a market, an economy in which even one's own death must be purchased. In the present, Dixie regards his own "body" as an abjected double—"this goddamn thing"—a regard that, in its very articulation, renders Dixie as a multiplicity, a being that is more than one.

This literary rendition of anticipation may seem to merely "represent" the desires of the computer-loving subjects of Silicon Valley for a tape backup of identity, much as *SimLife* author Ken Karakotslos describes his "hope" in the comic "About the Author" notes that come bundled with each game:

> Ken Karakotsios lives in Northern California with his wife Lucia and their three computers. He designed software in a previous life and hopes to some-day return to hardware, as a program.[23]

But such a reading of Karakotsios's "hope"—a symptom, no doubt, of the much discussed loathing of cyberpunks for "the body"—overlooks the productive work of such anticipation in the constitution of the subjects bundled with contemporary technoscience. "Uploading," the desire to be wetware, makes possible a new technology of the self, one fractured by the exteriority of the future. For Dixie, of course, this exteriority is figured as erasure, but what seems crucial to uploaded subjectivity is anticipation—a frenzied purgatory between the present and the future—rather than any specific destination. And while "anticipation" is, of course, an affect that has been available to hominids for some time, uploading seems to install discursive, material, and social mechanism for the anticipation of an *externalized* self, a techno-social mutation that is perhaps best characterized as a new capacity to be affected by, addicted to, the future.

In *Origins of the Modern Mind*, cognitive scientist Merlin Donald traces the relation between "external symbolic storage" and the cognitive evolution of humans. With the emergence of writing, Donald argues, human cognition became "permanently wedded" to external sources of memory in a "cognitive symbiosis unique in nature."[24] Donald argues persuasively that the evolutionary emergence of capacities for prediction and explanation—what he too simply characterizes as "theoretic thought"—were bundled with a dislocation of human memory. No longer localized "in" the human brain or body, memory becomes, with the emergence of writing, an external symbolic affair.

Without granting some of Donald's premises—the opposition between oral and written culture, the precedence of theoretical thought, the binary of a brain that resides "in here" versus a memory that resides "out there"—I would nonetheless suggest that his arguments concerning the complex ecology of technologies of inscription and human cognition are highly suggestive for uploading. For if external symbolic storage and theoretic culture rendered a subject with the capacities to be affected by the past—a complex of memory—then uploading technologies render a "subject" capable of being similarly affected by the future—a complex of anticipation. As a rhetorical practice and a technology of the self, uploading composes what Deleuze and Guattari have characterized as a "collective enunciation," a set of speech acts whose articulation only emerges out of a multiplicity—in this case the future(s). In contrast to the futural constitution of a Christian subjectivity, in which confession and other disciplines suture the interiority of a subject that encounters the contingent future—"Will I go to Heaven?"—uploading renders a subject entirely complicit with a contingency that is yet to come. As a kind of externalized brain or "exobrain," the uploaded subject launches synapses that remain fissured in the present and, perhaps, emerge as a thought of the future. The selves that emerge out of such a set of practices are defined by a relation not to a priest, nation, class, gender, race, or ethnicity, but to an enormous set of contingencies that can only be encountered rather than predicted.[25] Unlike the discourse circuit described by Foucault, such a technology of self is constitutively itinerant; it is precisely the singularity of futures that is being cultivated or encountered. "Terrifying to *whom?*"

While this may seem to be a literary affect provoked by an overdose on science fiction, new markets and forms of finance capital suggest that the frenzied encounter with a contingent future is constitutive of more than Gibson's characters or of Usenet discourse. For if the literary technology of the uploaded self still relies on a self as a "money changer" that examines itself through the lens of an economic transaction, the market that would render any such transaction intelligible has itself, sometimes, been uploaded into the computer softwares and financial instruments I discussed earlier: exotic derivatives.

As you may recall, exotic derivatives are extraordinarily complex financial instruments whose value can only be evaluated by computer softwares. A contract with the future, the exotic differs from the usual run of futures contracts in that it cannot be resold on a secondary market. Unlike, say, a contract on orange juice futures for 2002, an exotic derivative cannot be sold to another investor over the course of the contract. Instead, an exotic derivative is an algorithm whose *execution* is deferred into the future.

Many exotics are deployed as hedges, as instruments that smooth out the difference between the present and the future. So, for example, if AT&T is planning to expand into the Chinese market in 2006, it might purchase a contract that will lock in the cost of borrowing money at that time, freeze the price of fiber-optic cable and hedge the cost of computer chips. As such, the exotic forestalls the difference of the future—it is insurance against the contingency of markets yet to come.

But the *value* of such a contract cannot, by definition, be determined until the contract is executed. In terms of the value of such a financial instrument, then, one can only anticipate, based on the evaluation of the present or the simulation practices of a computer. Unlike a conventional futures contract, whose value can be determined by being sold, the value of a derivative resides in a strange intermezzo space of anticipation, a virtual haunting by the future.

Engines of Anticipation

This rhetorical production of anticipation—the smearing of the future into the very operations of the present—percolates through the quasi-popular, quasi-scientific discourses that uploading draws on and lives in. Indeed, in *If Uploads Come First*, an essay by researcher Robin Hanson whose title telegraphs the anticipated *jouissance* of uploaded replication, the anticipation of uploading is itself networked with the market:

> What does this all mean for you now? If you expect that you or people you care about might live to see an upload transition, you might want to start to teach yourself and your children some new habits. Learn to diversify your assets, so they are less at risk from a large drop in wages; invest in mutual funds, real estate, etc., and consider ways in which you might sell fractions of your future wages for other forms of wealth. If you can't so diversify, consider saving more.[26]

So too does the appropriately named Foresight Institute—one of the foremost resources for uploading discourse, founded by nanotechnological personality K. Eric Drexler—operate in such an anticipatory economy. Not merely a site for research into nanotechnology and uploading, Foresight's policy is "to prepare for nanotechnology" by

- promoting understanding of nanotechnology and its effects;
- informing the public and decision makers;
- developing an organizational base for addressing nanotechnology-related issues and communicating openly about them; and,
- actively pursuing beneficial outcomes of nanotechnology, including improved economic, social, and environmental conditions.

That is, the Foresight Institute is an institute of anticipation, a site of research into the effects of a technology that will, perhaps, emerge in the future. This program of anticipation emerges out of the writings of founder K. Eric Drexler. In his *Engines of Creation*, Drexler described the powerful effects wrought by a shift in anticipation or expectation:

> Expectations always shape actions. Our institutions and personal plans both reflect our expectation that all adults now living will die in mere decades. Consider how this belief inflames the urge to acquire, to ignore the future in pursuit of a fleeting pleasure. Consider how it blinds us to the future, and obscures the long-term benefits of cooperation. Erich Fromm writes: "If the individual lived five hundred or one thousand years, this clash (between his interests and those of society) might not exist or at least might be considerably reduced. He then might live and harvest with joy what he sowed in sorrow; the suffering of one historical period which will bear fruit in the next one could bear fruit for him too."[27]

Thus research into nanotechnology—and by extension, uploading—becomes a remedy for a blindness. This remedy opens a gaze to the future, a future whose technology of self will not "ignore" but will be constituted by events to come. Indeed, it is almost as if the first technology of nanotechnology is anticipation itself, a technology mobilized by Foresight. So too are the *effects* of nanotechnology cast as anticipation—the expectation of human subjects for extraordinarily long lives.[28]

An emphasis on the (often invisible) material economy of this anticipatory progam—rather than a meatless, transcendental realm where the body is "only" a rhetorical figure—highlights the contingency at play in the symbiotic encounter with silicon. Susan Oyama, writing of the metaphor of "program" within developmental biology, notes that while a program may be quite predictable (and hence not contingent) epistemologically, such algorithms retain an ontological contingency insofar as they are dependent on a Rube Goldberg economy of temporal and material factors for their instantiation:

> reliably repeated assemblies can be noncontingent in the sense of being highly predictable, while being thoroughly contingent in their dependence on complex, highly extended systems of interacting factors whose dynamic organization cannot be explained in terms of a single component or central agency.[29]

While Oyama writes of the operation of such contingency within biological systems, her principle applies equally to other instantiations of the program metaphor. A poster on sci.cryonics, in a discussion of uploading, put it this way:

> If we knew what the result of a computer program would be, there would be little point in writing or running it. In fact, it's a fundamental theorem of computer theory that in the general case, there's no way to tell what a program will do, other than to run it and see what happens.[30]

Crucial to this formulation is the constitutive role that contingency plays in "repeated assembly" of computer programs. The very "point" or purpose of the program involves, paradoxically enough, contingency. Becoming-silicon, uploaders *are* precisely such programs. Tied not to any single central agency (such as an autonomous self), programs find themselves contingent on the exterior assemblages with which they are implicated. Thus as a technology of self, uploading must be seen not simply as the erasure of contingency and the inscription of repetition. Instead, the dice throw of uploading highlights the necessity of contingency, an ontological and epistemological contingency called "the future" that emerges out of the rhetorical and material assemblage of uploading.

Within uploading discourse this anticipation of the materiality of simulation—not what software is, but what it is contingently capable of becoming—shatters the present, and disorients identity. In *If Uploads Come First*, the author notes that different moments of replication would yield different versions of identity in any given "present":

> Uploads who copy themselves at many different times would produce a zoo of identities of varying degrees of similarity to each other. Richer concepts of identity would be needed to deal with this zoo, and social custom and law would face many new questions, ranging from "Which copies do I send Christmas cards to?" to "Which copies should be punished for the crimes of any one of them?"[31]

Indeed, these "richer" concepts of identity provoked by the discourse on uploading include encounters with the effects that different material instantiations would have on each upload. Hans Moravec, one of the canonical authors of uploading discourse, notes the various dislocations of "identity" and "body" that would be fostered by different uploading platforms:

> If you found life on a neutron star and wished to make a field trip, you might devise a way to build a robot there of neutron stuff, then transmit your mind to it. Since nuclear reactions are about a million times quicker than chemical ones, the neutron-you might be able to think a million times faster.[32]

Of course, such a formulation preserves the category of a "you," an identity that is fundamentally the same, only faster, but the *mobility* anticipated by this argument is not simply a mobility away from "the" body; it characterizes a space where bodies are nodes in new becomings proper to humans:

> Fast uploads who want physical bodies that can keep up with their faster brains might use proportionally smaller bodies.... [A]n approx. 7 mm. tall human-shaped body could have a brain that fits in its brain cavity, keeps up with its approx. 260 times faster body motions, and consumes approx. 16 W of power. Such uploads would glow like Tinkerbell in air, or might live underwater to keep cool. Bigger slower bodies could run much cooler by using reversible computers...Other uploads may reject the familiar and aggressively explore the new possibilities. For such tiny uploads, gravity would seem much weaker, higher sound pitches would be needed, and visual resolution of ordinary light might decline (in both angular and intensity terms).[33]

Thus the anticipated body of the upload is not simply an occlusion or amputation of the body and its contingencies; it is a promised body, one summoned but not completed by the simulacrum. One poster to CryoNet, an enthusiast of both uploading and cryonics, outlines the debt that both such technologies of the self have to the contingency of the future:

> For me, the only thing I need to know is that there's a chance, there's hope that they might revive me—that's enough. As to whether I revive uploaded or meat, or even in Tipler's omega point, I don't care. My reason for considering cryonics is not to establish some kind of continuity of identity with the future; in fact I don't think that identity means anything useful. I think we're waves on the ocean. I think that "I" and "you" are just roles, like "left" and "right," and I don't invest those roles with any metaphysics beyond those of the context of consideration. Is "I yesterday" the same as "I tomorrow"? To me that's just tommyrot.[34]

While this may seem to be a late-twentieth-century retooling of Pascal's wager—what have we got to lose?—I would argue that a more profound affirmation of risk is at play here. Less enthralled to the immortality of the self than to the future itself, uploading becomes a technology of self that renders the univocal self "tommyrot." Indeed, for this poster, the desire to be uploaded resides not so much in the self as in the other of the future:

No, my reason for wanting to be suspended, or indeed for wanting to wake up tomorrow, is purely irrational. It's a creative urge. I like to build things, to write things, to nurture things and impart things to others. I don't care so much who receives those things, so long as they find them useful. . . . If my recipients are alive now, or if they're in the crew of misfits and lollygaggers that thaw me out in a few decades, or in a few thousand years, or at the end of the universe, I don't care—I find it equally satisfying to hope that one of these ways I might create more things.[35]

Less an ontology of loss than of anticipation, uploading sloughs off the body of the present and puts on the future, a future composed of contingency rather than eternity. In some sense, then, uploaders are dependent on, even addicted to, contingency. The possibilities of uploading are, through a feedback loop that remains to be mapped, entangled with the networks of technoscience and desire that would actualize its concepts. In addition, the futural practice of uploading is thoroughly networked with *other* uploads, even if those other uploads are replicants of oneself. That is, replication entails an iterability that would make the claim for the authenticity of any upload meaningless. For to be replicable, i.e., uploaded, is precisely to be iterable, and when it comes to iteration (as with some salty snacks), you can't have just one. As Deleuze and Guattari articulated it in *A Thousand Plateaus*, it is precisely this ability to be copied that is bundled with deterritorialization—"only something deterritorialized is capable of reproducing itself."[36]

Thus the unlimited character of the upload—its immortality dose—depends not upon its ability to render the "original," but upon connectivity, the capacity to be networked and even sampled by other uploads, a capacity available only in the future. Anticipation persists even through the actualization of the upload, as the structural possibility of an upload becomes the capacity for futures, a future not of persistence but of *sampling*, to become articulable in another moment or context, the context of silicon. As in Burroughs's method of the cut up, to become immortal is to continually become other, a continual variation. Adrift . . .

This becoming-sampled marks a new ecology of subjectivity, one in which the sheer exteriority and risk of the future—rather than the certainty of death—sculpts the subject of the present. The diagram of human cognition drawn by Donald and executed by multiple disciplines—an interiority constituted through an external matrix of memory, folded into a pocket of regret, death, and secrecy—itself becomes folded, twisted into a Möbius body that is neither past nor future, but a wetware intermezzo of anticipation.[37]

It will be objected that this new topology of subjectivity is not real, that in its dependence upon technologies that are yet to come, it is at best a messianic ethos. This is correct, if one persists in understanding uploading as an entirely technical project whose success can be determined in terms of actual uploads. But as a virtual entity—a concept—uploading is hardly immaterial, as it attracts resources—for research on nanotechnology, for example—and constitutes subjects in the present. Indeed, fusion with the computer could be seen to be the contemporary vanishing point of the human, an optical orientation that, like other vanishing points, organizes human life even though it is thoroughly fictive. I will close with an account not of a vanishing point, whose optics seem to have vanished, but with a regard for an encounter with the fracture, a fissure of consciousness that perhaps best maps becoming-silicon.

The Outcome; or, What You've Been Waiting For

Jean Baudrillard ends his now canonical text *Simulations* with a remarkable example of life in the field of simulacra, an example that perhaps encapsulates the all-too-speedy-and-greedy nature of Baudrillard's formulation of simulacra as the "desert of the real." Speaking of the conflation of the real and its models, a moment in which the real becomes nothing other than the ability to *be reproduced*, Baudrillard meditates on a moment when

> each is already technically in possession of the instantaneous reproduction of his own life, where the pilots of Tupolev that crashed at Bourget could see themselves die on their own camera.[38]

Whereas Baudrillard, in a kind of instant metaphysics, a metaphysics of the instant, collapses the execution of a death with its simulation, this very example articulates the way in which uploads enthrall through anticipation. The witnessing of one's own death, as a virtual event, can only be rhetorically and not corporeally rendered. Rather than enforcing Baudrillard's declaration that the real has disappeared in favor of the implosions of the hyperreal, the gaze of the pilot allegorizes the *futural* character of uploads: such replicants enthrall with the impossible, next image, a becoming-visible, an agonizingly gorgeous unfolding.

Gilles Deleuze, in his two-volume work on cinema, offers a compelling diagram of such an image. As I discussed in the case of the dog, Lazarus, Deleuze argues that while the classical cinematic image operated through the visualization of movement, the mutant sign of the postwar cinema is the time-image. Whereas the movement-image, Deleuze claims, animates movement through an absence—a hole or gap between cinematic frames "where" movement occurs—and focuses on the apparently self-moving entities within frames, the time-image renders the gap or cut itself, the "intermezzo" or the "interstice." If the classical cinematic image operates with a movement "buried" in the interior of the cinematic assemblage, a visible invisibility of the present, then simulation, particularly uploading, seems to work through a becoming-visible of the future, a becoming whose rhetorical effect is "anticipation" and whose "ontology" is algorithmic and complex. Rather than composing its image through gaps in the present, the simulation is constituted out of a fracture or a fractal, the unlimited finitude of the future, virtuality crashing into actualization.

What You Were Waiting For, Part Two: Long Live the New Flesh

David Cronenberg, in *Videodrome*, meditates on a kind of "uploading" onto video.[39] At the end of the film, when the main character, played by James Woods, raises a newly grafted and thoroughly oozing gun to his head as a sacrifice of his old body in favor of the "new flesh" of video, we are again, like the pilots at Orly, anticipating the virtual witnessing of one's own death—Woods is, of course, on video as he makes the transition to the new flesh *of video*. But the anticipatory gaze leads to a blank screen, an interstice that paradoxically comes at the end.

Rather than a simple refusal to display the "world" of the new flesh, Cronenberg's production of a blank yields the mechanism of the "new flesh," an invisible mechanism whose visibility is continually anticipated but which is imaged only through a fracture or a break. Robin Hanson's *If Uploads Come First* similarly figures this image of the future in his subtitle: "The Crack of a Future Dawn." Uploading is thus an anticipation of precisely "more life," life not free of the body but distributed into spaces not yet visible, our best name left for the remains of life in the age of the simulacrum, the algorithmically complex whose only ontology resides in its instantiation, an instantiation incessantly at play in the future.

Uploading, it is clear, is a technology of self that as yet replicates little. It intrigues not so much because this technology of self is common or even likely to be so. Rather, uploading discourse offers a map of contemporary technologies of self, technologies that are contingently sculpted. Addicted to contingency, or prowling for mastery, uploaded subjects mark out the materiality and possibility of the new flesh, anticipated flesh that is something other than either transcendental or meatless. As for my own anticipations, I'll adopt an affect not of anxious waiting but of cheerful and prankish hailings of the future. "More contingency, fucker," or to sample Alluquere Roseanne Stone in a rather anticipatory mood, "I hope to observe the outcome."[40]

S E V E N

Dot Coma:

The Dead Zone of Media and

the Replication of Family Values

They'll get a fuckin shock, when they see this near-corpse, this package of wasting flesh
and bone just rise and say...

Irvine Welsh, *Marabou Stork Nightmares*

I have argued incessantly throughout this growth that the massive changes wrought
by the narratives and practices of molecular biology have shifted the very concept of
life at play in contemporary culture as distinctions between living systems and machines have begun to blur and morph.[1] No longer attached to organisms, life becomes
an emergent attribute of information systems, networks without any obvious center.
In the example of artificial life, contemporary culture is beginning to be populated
with entities whose "life" is both uncertain and difficult to locate. Uploaders form a
futures market for a subjectivity franchise on the Internet domain, an ex-corporation
whose installation shatters the autonomous individual into a sample of continual
variation. More than a repetitious immortality, uploaders provoke a new relation to
the future: being-sampled. The very transformations that would make possible the
copying of human subjectivity onto silicon also enable new forms of deterritorialization, the promiscuous splicing of subject effects not unlike those acts of bacterial
conjugation, transduction, and transformation detailed by biologist Lynn Margulis.[2]

These new distributions of vitality—both fantastic and scientific—are not confined to that usual ecology of virtuality, silicon. Comatose bodies cultivate yet another, singular execution of an informatic body. Accompanied by more than the visualization of an EEG and the machinic yoga of life support—Breathe in! Breathe out!—coma patients are connected to multiple rhetorical machines that would govern this strange flesh and enable its narration. In what follows I map out some of the capacities and threats posed by those bodies whose vitality is articulated as a signal, "wetwares" through which contemporary informatics are instantiated.[3]

Out of the Barrel of a Gun

Gary Dockery of the Chattanooga Police Department was shot in the head in September of 1988. He subsequently went into a comatose state, and was silent for seven years. In February of 1996, Dockery developed pneumonia, and his condition worsened as his lungs filled with fluid. "His family was given the choice of risky surgery or letting the disease take his life." Such a decision, of course, proved difficult: How to manage the enormous contingencies associated with such a calculation? On the one hand, the family had good reason to wonder if the pneumonia was not a blessing, an end to the lengthy suffering he had endured. At the same time, the very indeterminacy of Dockery's state argued for further aid; some argued that he had suffered enough, others thought that no help should be spared.

Only the agon and difference of argument could broker such a decision; a true differend, the Dockery's decision could not be made through recourse to any maxim or law, even as it was the law itself that distributed the decision to the family. Family members argued the situation in Dockery's room as Dockery grappled with pneumonia in silence.

Four hours later, Dockery spoke. For eighteen straight hours, Dockery spoke of the time, read thermometers both digital and analog, and said that he "did not want to go back to the village," a nursing home facility where he had been "living." The sheer excess of the discourse, as well as its novelty, provoked media interest, and within hours the story was reported in print, television, radio, and web sources. The volume of talk about Gary Dockery's coma, both by Dockery and the media, would be difficult to quantify, but it unerringly focused on the odd rhetorical situation the Dockerys found themselves in. Of what caused Gary's return,

> Dr. Folkening said Gary Dockery's illness, the change of environment, an onslaught of visitors after years of sparse contact and hearing discussions of his death may have contributed to his awakening.[4]

In light of Dockery's discursive eruption, the family decided to go ahead with the surgery.

But even as Dockery's speeches would seem to eradicate the ambiguity of the coma—clearly he was alive, unambiguously responding to his name, talking with his child on the telephone—a new and disturbing uncertainty emerged out of Dockery's clarification. If Dockery had made it clear to his family that he indeed wanted to live, his very persuasiveness threatened the fragile certainty that surrounded comatose patients everywhere. If such agency persisted in the coma victim, a patient for whom families must speak, it would render even more difficult and uncertain the calculations that undergird life support. Even after his testimony, the Dockerys were unsure of the decision to operate, as the surgery could once again plunge Gary into silence even as he spoke for his life for eighteen hours. On the Christian Broadcast Network, Gary's mother spoke of their discord:

> I had to wrestle with some of them. Some of them did want to give up. As a matter of fact, last Sunday most all the family, his son and even some of my children wanted to give up and just go ahead and let him go, and I refused because God gave me a promise.[5]

But the ambiguity provoked by Gary Dockery's blue streak was not confined to the decision regarding his own surgery; it would haunt the deployment of decisions made by families everywhere. These decisions were already troubled by the impossible evaluation of the contingency or worth of the patient's life; now the dangerous outbreak of comatose agency in Chattanooga undermined the worth of familial opinion itself; Dockery's voice would seem to have more weight than the legally legitimated voice of the family, that voice through which the coma speaks. So too was the popular image of medicine under attack:

> The wonder of the Gary Dockery saga is why did he speak just as his family was resigned to letting him die?
>
> "We have no explanation at this time," Dr. Kaplan says.[6]

The silence of the medical community only exacerbated the import of a prior silence—that of comatose bodies:

> I think the biggest interest this will spark is mutism after this sort of injury. We assume mute means that not much is going on. The fact remains that Gary Dockery was mute for seven and one half years and now is capable of speech.[7]

The medical profession worked quickly to dispel any fears that the brain dead were anything other than unambiguously dead. While early reports of the Dockery case spoke of his emergence from a coma, not "the mummy walks" but the coma speaks, physicians gradually pointed out a subtle distinction between two different types of brain death—the persistent vegetative state (PVS) and whole brain death or brain-stem death. On National Public Radio we learned that

> Dr. John Caronna, a professor of clinical neurology, tells Noah that the story of Gary Dockery's waking up from a 7-year coma is not entirely accurate. Medically, Dockery has maintained consciousness, but severe brain damage from a gunshot wound limited his response to stimuli. Caronna says something energized him, increasing his ability to communicate. But it's unclear if he will continue to improve or not.[8]

This "grandfathering" of Dockery into a PVS rather than a whole brain or brainstem death failed to eradicate the ambiguity that suffuses the coma patient. How could there be more than one type of brain death? The proliferation of distinctions makes possible finer grades of discipline on the comatose body—indeed, it makes the difference between being a "neomort," available for organ donation, and being a comatose subject, available, contingently, for the future—but it also undermines the univocality and persuasiveness of the declarations of death. No longer a binary "She's alive!" or "I'm sorry, we lost her," the diagnosis of death now becomes a continuum, with each distinction threatening to blur into the next, while the comatose body prepares to be "energized." The "something" that enabled a comatose body to speak in this instance was, of course, a very specific species of speech act or virtual witnessing that operates through the absence not of the other but of the self: "hearing discussions of his death."[9]

Of course, "comas" have always been speech acts. First articulated as a medical definition by French physicians Mollaret and Goulon in 1959, the *coma depasse* was redescribed by an ad hoc committee at Harvard University as "brain death" in 1968.[10] By 1981, the new taxonomical category of "brain death" became part of UDDA, the Uniform Declaration of Death Act. The twenty-two-year period that spans Mollaret and Goulon's discussion of *coma depasse*, irreversible coma, and the retooled "declaration" of death testifies not only to the distinctive shift that had taken place in the constitution of legal death, from corporeal movement to televisual signal. It also indexes the heterogeneity and contingency at play in the various criteria deployed around these boundary criteria, a boundary that is seemingly self-evident but remarkably murky: life/death.

This ambiguity is not, of course, confined to the present. Physician John Cheyne, in his classic 1812 text *Cases of Apoplexy and Lethargy: With Observations upon the Comatose Diseases*, finds paradoxically that apoplexy (the coma's genealogical precursor) strikes those who seemed least likely to be struck down by ill health.[11] And Cheyne's attempts to define "apoplexy" meet with only what he would characterize as a "rhetorical" satisfaction.[12]

The continual ambiguity and contingency that haunt this "declaration" of death is evinced in numerous accounts of medical practice. Bioethicist Lance K. Stell writes of the effects of death's apparent new multiplicity:

> Despite the intent to underscore the neurological basis of traditional criteria for death, the term "brain-death" has itself exacerbated confusion. To many laypersons (and to some medical professionals too, unfortunately), "brain-death" suggests that there is more than one kind of death ("brain-death" and "cardio-respiratory death"), or that there is more than one way to be dead (in a brain-sort-of-way and in a heart-sort-of-way), or that there are degrees of being dead ("brain-dead" and "really dead" or "dead-dead"), or that one might die more than once (first, when one's brain dies and again later when one's heart stops).[13]

This continual multiplicity provokes the question that I ask as a rhetorician: How is this ambiguity and contingency managed in the narratives and articulations of the comatose body today, and what sorts of bodies and configurations of power do these management tactics enable?

The outbreak of excitement and uncertainty at the site of Dockery's comatose body was managed, by the media, through relentless recourse to Dockery's family. The media, too, talked a blue streak, and the unnerving possibility that it may have been the family's discussion that provoked the outburst was continually alluded to even if it was rarely named. So too did media reports imply that Dockery's incommunicado was attributable to a lack of attention on the family's part. In the first year after the shooting, Dockery communicated with blinks, nods, and grimaces, but "that stopped when visitors diminished after a few years."[14] At the same time, the family itself became the vector of comatose reanimation, as it was the very voices of the family that were presumed to have "energized" Dockery.[15]

Ungovernable even (or perhaps especially) by a medicine that deploys finer and finer distinctions, the comatose body becomes the promise and burden of families, an obligation toward the proper comportment of a body and its organs toward a future riddled with contingency and subject to the continual articulations of a spectacle. While Dockery's mother "wrestled" with family members in the incalculable arguments over his surgery and his future, she also had to wrestle with the ongoing interest of the media. Indeed, according to the Web page devoted to the Dockery family,

The Reunion Will Welcome Two Special Cousins

Two Special Cousins plan to attend the Association Reunion in Cherokee Co., NC the second weekend of September,1996. They are:

Dennis Dockery of Chattanooga, TN, who is a brother of Gary Dockery. Gary was the policeman who was shot in the head while on duty and had been in a coma for seven years when he suddenly began to talk this past February. This miracle story was widely reported by news media across America. Dennis will bring us up-to-date on Gary's condition, as well as how the family has attempted to deal with the media blitz.[16]

My point here is not that families should not be the locus of power in the complex economy of medicine and culture that surrounds the comatose patient. Rather, this medical and media episode highlights the rather unstable character of the comatose body, a body for whom no medical distinctions appear adequate and whose diagnosis is rife with ambiguity. In this situation it is the "values" of the patient's family that are called on to legitimate the governance of the comatose subject, "values" that also find little strength in distinctions. Rather than a sign of contemporary medicine's humanism, its tender yielding in the face of familial wishes, the recourse to the family as a site from which to govern the comatose body underscores the familial disciplines that form the unspoken ground of contemporary health, health that ultimately fails or returns under the aegis of familial "care." If health is life lived in the silence of the organs, then the coma is a silent life sustained within the noise of familial discourse.

A Phone Is Ringing...

Hello?

A telephone was ringing, somewhere else. A voice *interrupts* the ring, somewhere in the middle, between the iteration of one ring, followed by another, and another.

The voice rings true. It is indeed who we think it is. A dead mother, separated from her son by war, by the Nazis, those horrendous assemblages of division and erasure, *lives*, and the formerly infinite distance of death is bridged by an area code, an exchange, a four-digit number. We get death's number—a simple inquiry into "information" suffices—and in return we receive a word, indeed, an "address."

"Hello?"

This speech act—one that hums with an analysis provided by Avital Ronell, on another line—comes as a surprise, like all interruptions.[17] It is an eruption of a question. Noisy, this unprecedented speech act sounds difference, the difference of an altogether other future: a future of life is on the line.

But we have to hang up, because this is the movies, a cinematic deployment of comas and their effects that, perhaps, doesn't *speak* to us. Our film, in hanging up, interrupts itself, images interruption. What had been promised here, in David Cronenberg's adaptation of Stephen King's *The Dead Zone*, was the visual rendering of a comatose body that would, somehow, tell us its future, a future that does not, unlike the coma, bear the burdens of contingency.[18] Cronenberg's camera images John Smith, a high school English teacher (Christopher Walken) smitten with the writing of horror—"Tomorrow, we'll discuss the legend of Sleepy Hollow. I think you'll like it. It's about a school teacher who gets chased by a headless demon"— as a coma victim who experiences a kind of second sight upon his awakening from a five-year coma, a period of no time at all for Smith (or the viewer) but which spells five years in cinema time, thirty-five years to a TV dog. This second sight is less a mastery of the exercise of prophecy than an overtaking, a wave of the future that impacts Smith with all the terror of a force from the outside—the exteriority of thought, the "outside" of the body, the materiality of time.

Overtaken by an image of the future, Smith acts to intervene in the image, to act in it and on it. Hacking the future, Smith attracts familial connections—they are the point of articulation for this second sight. The coma victim's psychic powers are invested at nodes in the familial network. "Your daughter," he screams to a nurse, "is screaming." "Your mother," Smith tells his astonished doctor, "she's still alive." Family, in this rendition, becomes the obligatory passage point for any encounter with the future whatsoever. Family, and its repetition of the tired old laws of lack, remains the hang-up that will not be interrupted.

The potency of the family—its unique character as a site that can channel, in effect, the future—is not materially limited to Johnny the coma victim's strange powers. It also emerges as a second-order effect, a virtual ricochet of an actual vision. Johnny's deployment of his second sight under the media glare of television cameras provokes a stroke in his mother as she views the screen. The knowledge of a familial secret—"You want to know why your sister killed herself?" Johnny says to an overly inquisitive member of the media—operates on maternal flesh through the virtual relay of the screen, at a distance. Nuclear holocaust—one version of the future rendered in Smith's second sight—is avoided when a populist, neofascist political figure takes refuge from an assassin by hoisting the body of an infant to ward off bullets, bullets fired by Smith in his coma-augmented second sight. The flesh of heterosexual reproduction literally armors the present and forestalls the future, in all of its terrible difference.

Perhaps this investment of the power of the coma in the familial network is incidental to its representation in Cronenberg's vision. But bundled with contemporary renderings of the comatose body as they appear in diverse American media outlets and international ecologies of literature, it would appear otherwise.[19] Instead, comatose bodies and subjects are incessantly articulated through familial dramas, family "units" that are invested with new powers even as capital disperses, distributes, and networks the nation-state. Ungovernable by states—who can only excel in the melancholy of *waiting*—it is only through the family that a coma patient "speaks."[20]

Waiting, of course, is a practice of the interval, a space between actions. The example with which we were interrupted—the dead zone—nicely contains the various ambiguities of this space "between" life and death by eliding them. Smith awakens, scarless, and sees his healed body as a miraculous sign—all smashed up but "no bandages." Similarly, the trauma that the comatose state poses for narrative—What will happen next?—is rendered invisible, passed over in the traversal of one scene by another. And yet the radical ambiguity of the comatose state—Is he alive? Is he dead? How should we comport ourselves toward him?—produces other effects, the dissolution of the opposition between past and future. As the mystery of the coma is figured here precisely through its elision of time—"You gotta understand," Johnny explains to his former fiancée, "For me it is like we just spoke yesterday"—it is through his uncanny existence outside of time—the dead zone—that the coma continues to do its work in the narrative.

But even as such an image arrests the ambiguity of the comatose body—a body that, in its return, can see outside of time, obliterating contingency—the alterity that seems to inhere in comatose bodies returns. The very source of Smith's second sight becomes undecidable, as soon as the stupor of the cinematic experience is shrugged off. The paranoid viewer recalls that our schoolteacher had effectively prophesied his own coma with his assignment—"The Legend of Sleepy Hollow." The possibility that the prophecies deployed by Smith are of a literary and not a comatose origin haunts even this fantastic management of the thoroughly ambiguous coma, a state of corporeality that smears our definitions of life and death. Under this reading—an experience of the film that one cannot *choose* but whose effects are irreversible—the comatose interval becomes less a cause than a trigger, a catalyst for the second sight whose roots are, very possibly, literary.[21]

As viewers of such a film, we are then placed in a remarkable quandary, even a hollow. Certainly, one could merely shrug off "Sleepy Hollow" as a cheap foreshadowing, a minor detail that retroactively produces a sense of "aha." But such an experience—the momentary, *jouissance*-laden gasp of recognition—is itself isomorphic to the profound interruption figured by the coma. That is, such an "aha" itself functions as what writer Catherine Clément has characterized as a "syncope," a kind of micro-awakening that proceeds from a micro-coma, a forgetting provoked by the viewing of a film, the reading of a text.[22]

Thus even the degradation of the ridiculously literal B-movie foreshadowing fostered by "Sleepy Hollow" serves only to provoke yet another question: *Where have I been?* While the viewer is hardly in the situation of corporeal injury referenced by Christopher Walken's character—all smashed up—she nevertheless enjoys an isomorphic dislocation. Some small element of the past returns, and is recognized as a prophecy of the future, a telltale sign not heeded until it is literally too late—the present. Such a dislocation puns on Johnny's affliction—it is the second "cite," the iteration of a text or an image through which one encounters one's place "between" two cites, a hiatus or a breath through which the "aha" emerges.

Cronenberg's and King's rewriting and repetition of the Sleepy Hollow tale—itself a tale of the terrible effects of the literary—provokes the limbo of the coma in the viewers of *The Dead Zone* even as it ensures that the comatose body will only speak through the family. The rhetorical algorithm of *The Dead Zone* operates through an acceleration of the very logic that makes film possible. The interruption—that moment between Smith's entry into a coma and his emergence—injects Smith with precisely the ingredient needed to "see" the future—interruption itself. Not simply a lack or down time, Smith's hiatus becomes a "sleepy hollow," a space between life and death, one frame and the next, an interstice that enables the invisibility par excellence—time—to be imaged, transformed into a "zone."[23]

But what is imaged in *The Dead Zone* is not so much the future "itself" as the *character* of futures, events whose arrivals are syncopated in a rhythm of continual interruption. Interruption—that communication breakdown—is here figured as what the future *does*. Smith does not *know* the future—he doesn't even know, in the present, what the status of his encounter with the future *is*. Instead, like the syncope that destroys his mother's body, the future interrupts Smith, and he, unlike our first caller, cannot hang up.

Such a logic of interruption—to encounter the future, let it break you up, put you into a coma, kill you—cannot, like our reading, be chosen. Such a procedure is articulated in Deleuze and Guattari as the "connect-I-cut," recipes for becoming that depend on a healthy dose of forgetting and breakdown.[24] As with exotic derivatives—those morphologies of capital whose encounter with the future depends crucially on an interval of nonknowledge, a dead zone of value—the comatose body is constituted by its noisy, inarticulable silence: what will happen?[25] In this register, *The Dead Zone* maps the governance of the comatose body by making it speak, and in so doing outlines contemporary modes of subjection—literally, the emergence of comatose individuals as they are networked with the family. The family ventriloquizes the subject as a crucial aspect of its own propagation, a distribution that is not simply of a biological order.

Comatose bodies challenge the referential capacities of discourse. As in the fetus, they provoke a referential panic, a *pro-lifer-* ation of attempts to render a sturdy border between life and its others.[26] Sheer growth without consciousness, a becoming-plant, comatose bodies are uncanny for their doubling of life. One sees the signs of life, one is even drawn to speak. "Don't leave me Johnny. We're gonna get married." Yet in the end, one just doesn't know the nature of one's audience— where does the machine end, flesh begin?

But comas threaten more than the boundary between life and death or flesh and machines; they disturb reproduction itself, or at least the family's monopoly on the propagation of human life through heterosexual reproduction. After awakening from his accident—a close encounter with a *milk truck*, the beginning of a machinic nursing that allows him to wait for the future—Johnny is visited by his parents. "You've been lost for five years, and now reborn unto me," his mother declares. With her biblical citing, Vera recaptures the strange eruption of life out of the machinic environment of the coma into the logic of reproduction—"reborn unto me"—and reattaches the generation of vitality to a mother. "Lost," without reference to the family, Johnny's miraculous rebirth from the machinic labor or "trance" of the coma is the second cite or site of birth. In naming it *as a birth*, Johnny's mother both marks the threat to heterosexual reproduction posed by his revival and manages it through recourse to the maternal body.

For if life can emerge out of multiple connections to machines, the role of heterosexuality in the propagation of a human future becomes visibly and disturbingly questioned. In place of the alliance of ovum, sperm, and futurity, Johnny's birth marks the new capacities of machines. And unlike the offspring of reproduction, on whom the mark or "navel" of originality persists, the progeny of a machinic economy of replication foster an erasure of reference, a doubling that is at the core of the notion of immortality associated with cloning and uploading, a repetition that sutures the very interval foregrounded in *The Dead Zone*. In place of the astonishing arrival of novelty associated with birth, such replicas offer the "already seen" or second sight of *déjà vu*: "There I am again."

Indeed, Johnny's "accident" all too predictably occurs precisely through an interruption of heterosexual reproduction. Instead of spending the night coupling with his fiancée, Smith ventures out to drive on roads slick with rain, promising to marry Sarah and noting that "Some things are worth waiting for." In the accident—itself caused by a trucker's sleepy, nodding hiatus—much blood, and much more milk, is spilled.

During the five-year wait—what in banking terms we might call the "float" of Johnny's promise to marry—Sarah, in the words of Johnny's mother, "cleaves now unto another man, a husband." To "cleave," of course, means both to split and to attach, a connection that puns on cutting. In this instance, Sarah's connection to the husband produces a child, a ten-month-old son she introduces to Johnny with Freud's famous epithet—"his majesty." Sitting around the table with Johnny, Sarah, and her son Denny, Johnny's father remarks, "You know, it feels good to have a family eating around the table again."

This network of examples seem to connect family—whether in its simulated form at the dinner table, the sovereign kingdom of the baby, or in the actual flesh of Johnny's mother—to any means of making the coma refer. In the discourse of his mother, Johnny's emergence from the coma marks his hiatus as a long labor, a labor that in the end is a rebirth unto the maternal body. In place of a lively ecology of machines, Johnny is reborn unto a mother, ensuring at least a rhetorical monopoly of the maternal on reproduction.

In the context of the simulated family gathered around the table, the coma is occluded through recourse to the evidence of Denny, a child testifying to the apparent fantasy of the past—"his majesty" the king whose word, or silence, is law: the coma never happened, Johnny and Sarah's coupling was never interrupted, and the maternal body of Johnny's mother has been exchanged for the reproductive success of Sarah. Here the family form is established precisely to the extent that maternity and paternity—those signs of "family"—are iterable and thus detachable from any particular context. In this milieu, the immortality of the family form itself—its ability to replicate over time, ad infinitum—helps manage the strange contingency and finitude of the coma.

To make sense both of the odd state of the coma and the miracle of its closure, the eruption of life out of machines, families are mobilized. In the face of what Félix Guattari has characterized as new "collective arrangements," social and technical groupings that would bestow and maintain the life of human beings, narratives of the coma recuperate the monopoly of the family on life by making the coma speak only on the register of the familial. Indeed, even in his flight from the traces of family—he no longer can bear seeing his former fiancée—Johnny provides us with his second cite from Sleepy Hollow: "As he was a bachelor, and in nobody's debt, nobody troubled their head about him anymore." An alleged escape from the familial matrix operates only through the double negations of family and capital—bachelorhood and indebtedness. The "value" of the family, in this context, emerges from an ability to evaluate the strange event called the coma.

In contrast to the comatose body, which miraculously emerges from an environment of machines and care, his majesty heterosexual reproduction is conceptualized—by a Christianity of the father, a psychoanalysis of the Law, and by that state privilege dubbed the "tax write-off" in the contemporary United States—as a Euclidean, triangular relation, Mommy/Daddy/Child. This clean logic—one which continually reinscribes the "autonomy" of the fetus/child, the father, the mother, as if all weren't entangled with all—tames the obvious multiplicity of heterosexual re-production into an order: heredity. Reproduction's territory, its necessary habitat, becomes narratable in terms of a name. As such, family operates as an algorithm, a recipe for converting the syntactic distributions of alterity—the chance and "drift" at play in evolutionary systems—into the semantic, narratable regimes of a people that know who they are: "You are a Doyle!"

This command at the heart of the family is threatened by the comatose body's refusal to signify—"I can't hear you!" as a drill sergeant might put it. It is disturbed as well by the visible rhizome of connections that make the ongoing life of a comatose body possible, a tangle of connections that renders laughable any simple demarcation of the comatose "subject."[27] Mute, without response, comatose bodies seem always about to speak. The strange confessionals that accompany the rhythmic, machinic cadence of the coma are judgments that the coma is a becoming, a trajectory and not a state. About to speak, about to die, the comatose body fosters continual becomings-other.

The fact that something else, it seems, is always about to happen to the comatose body foregrounds the incessant labor and discipline the familial franchise demands, as this anticipatory state calls forth every interpretation of every sign that can be mustered. This discipline—the means by which families might maintain their monopoly on reproduction, on human life itself—operates not only through the deployment of flesh, but through the propagation and actualization of concepts. In the case of comas, an uncanny entity I will call the "virtual maternal" forestalls the possibility that something other than human, heterosexual reproduction is entailed in the emergence of a coma survivor's life.

And Mommy's on the Phone...Or, "It's a Coma!"

Feminist scholars have argued convincingly that the threats posed to the familial monopoly by reproductive technologies have recoiled onto women's bodies. The rhetorical amputation of the fetus from maternal bodies bites into actual flesh as women and uteruses have increasingly become incarcerated, surveyed, purchased, and disciplined. But the new technologies of life *maintenance* also interpolate the contemporary family, and this threat entails precisely the obliteration of the body enacted in the discourse of fetal "rights." In this case, however, the maternal body is far from invisible—it becomes a privileged site for making the comatose body speak, a body that the family recomposes out of its fetal obliteration. This composition operates not through the attachment of a fetal body to actual flesh—in *The Dead Zone*, mothers are either killed, telephonically terminated, or shot—but through the rhetorical production of maternal effects. Having liquidated the body and agency of motherhood in the contemporary United States, families recompose the maternal body, the possibility of heterosexual reproduction, through virtual tactics. Forced to speak out of her silent labor, the mother appears in *The Dead Zone* as a telephonic entity, one who mothers without consciousness, a machine for reproduction.[28]

If a comatose body's trajectory—its becoming—is rendered unto narrative by the maternal body, this rendering is an uneasy, even labored, one. Ventriloquizing biblical prose, Johnny's mother, Vera, attempts to connect his return to the divine matrix of reproduction. But the very oddity of Vera's speech—its character as a *cite*, a machinic repetition from elsewhere—unhinges the connection: Neither resurrection nor birth, Johnny's return appears without causation, without origin, a "miracle." Nursed by networks of machines and care, the comatose body puts a new spin on the reproductive concept of "expecting."

For if Smith—or, as his "John Doe"-like name suggests, any coma victim whatsoever—emerges into life out of a machinic environment, it is a life of the future. Befallen by the future, the comatose body is a virtual body, one continually anticipated but not yet actualized. One speaks to it and waits: "waiting patiently for something to happen."[29] Literally unconscious, alive, comatose bodies become conscious subjects through the continually, hopefully expected interruption called the future. Without recourse to consciousness—of a mother and her labor, of a fertility clinic, of a surrogate—life emerges, veritably laboring with the future.

Valerie Hartouni articulates this odd capacity of the comatose body to labor in her 1991 essay, "Containing Women: Reproductive Discourse in the 1980s." Hartouni focuses on the ability of the maternal body to disappear under the increased visibility and agency of the fetus. But Hartouni's analysis also enables an understanding not of the disappearance of the maternal but of its deterritorialization, its becoming-virtual. Hartouni describes the case of a newspaper headline: "Brain-Dead Mother Has Her Baby." In this instance of a comatose body that has become downright swollen with agency—for clearly in this instance "having" is most definitely a "doing"—Hartouni highlights the paradoxical annihilation of maternal agency as the "mother" becomes less verb than noun. In the context of its citation into a headline, Hartouni notes that

> motherhood is equated with pregnancy and thereby reduced to a physiological function, a biologically rooted, passive—indeed in this case, literally mindless—state of being.[30]

Indeed, Hartouni argues that, as such nouns or states of being, pregnant women are merely "mediums or physical vessels for new life, not active participants in its creation or maintenance."[31]

But what are the capacities of "mediums"? Hartouni's analysis suggests that media are bereft, impoverished in agency—"passive"—and immobile—"biologically rooted." But perhaps a medium—like Johnny in *The Dead Zone*—is precisely in the middle, neither active nor passive, present nor absent, live nor dead, here nor there.[32] Perhaps media are virtual, on the line. As virtuals, such media do not lack agency but instead distribute it—they articulate the connections in any network of actualization. As such, a medium's promises and threats emerge from their location in limbo, as go-betweens.

This limbo agency of the virtual resides partially in its capacity for repetition. At times, Deleuze and Guattari describe the very distinction between the virtual and the actual in terms of speed and slowness, as actuality arrives in a kind of interruptive "freeze frame" that brings reference to bear on the multiplicity of the virtual. These differential speeds can be understood as connectivity, the facility to be repeated in different contexts, an attribute roughly analogous to the quality of "velocity" in monetarist economics. Betweenness has its benefits—connectivity increases with deterritorialization, while the valence of connection increases until it undergoes a change in kind—reference is qualitatively other than persuasion, actuality differential of virtuality.

This transformation of a virtual into an actual requires repetition and it is through iterability (and subsequent distributivity) that virtuals are "selected." For even Hartouni's deployment of the example/sample "Brain-Dead Mother Has Her Baby" relies on its extraction and repetition. No less than the delivery of the fetus/baby itself, Hartouni's mobilization of "Brain-Dead Mother Has Her Baby" relies on its ability to be cut out of its alleged "location," much as Hartouni notes that with the rhetorical operations of video and visualization technology "the live fetal image of the clinic appears to have been transported into everyone's living room."[33] Indeed, in this sense, actuality—the emergence of reference—is thoroughly entangled with virtuality—the capacity for dispersal. This is stated most clearly by Deleuze and Guattari when they note that "there is no reproduction without genetic drift."

Such a power of drift, distribution, or delivery is indeed an attribute of media—they function as speedy replicators as well as articulators and are sometimes contagious enough to sprout, ungovernably, in multiple contexts. More than an accidental predicate, media entail a primordial dislocation at play with the real.

> The virtual must be defined as strictly a part of the real object—as though the object had one part of itself in the virtual into which it plunged as though into an objective dimension.[34]

In this instance, as in every other, only by being treated in terms of what Deleuze calls differentiation—a partitioning into articulable, narratable distinctions, "as though"—can the virtual be rendered. It must be actualized *as a virtual*, determined in configurations of space and time that enable it to refer, to become a "problem" for which actualization, becoming, is the solution. It is through the virtual that the monstrosity of metamorphosis can be referenced, treated not as a vague promise—of the tadpole, "maybe it will be a frog"—but as the positivity of becoming. If virtuals form a hollow of an "objective dimension" where the real has its tendrils, they are no less complicit with their own delivery by the future, the capacity to be acted upon. Deleuze describes the differential transformation of virtual into actual through recourse to the figure of labor—at first of the economic kind, but then through the embryo:

The destiny and achievement of the embryo is to live the unlivable, to sustain forced movements of a scope which would break any skeleton or tear ligaments.[35]

This logically impossible event—living the "unlivable"—proceeds not through the usual operations of "agency," "action," or "strength." It instead requires a flexibility, a tremendous facility for difference and a primordial hospitality—egg, nest, hollow. This sensitivity to difference operates in contact with the future: "They can only be lived..."[36] It is in this sense that even the virtual is complicit with the future, and that mediums enable and distribute agency more than they suffer its lack.

Hence Hartouni's understanding of the "passive" "biological" mother must be supplemented with a mapping of the effects of the very "mediums" she fears are being evacuated of agency. It is exactly the flexibility of mediums—their ability to be hollowed out, like Johnny, by the future—that makes the arrival of novelty, indeed birth, possible. In comparing possible alternative formulations of "Brain-Dead Mother Has Her Baby," Hartouni notes that it is through the disavowal of the "social activities and meanings on the one hand" of the mother and the amplification of "biological processes on the other" that the comatose mother is constituted as a medium. And yet the event of the technologically and rhetorically entangled "Brain-Dead Mother" is exactly what draws these distinctions into question *as oppositions*, suggesting that reproduction has perhaps sprouted more than two hands.[37]

That's One Way of Putting It; or, Aha

Dependent on an unforeseeable future, the emergence from a coma is also, of course, difficult to narrate. Upon awakening, Smith's doctor informs him that he had been a "guest" for some time, a locution Smith greets with a snort while responding, "That's one way of putting it." This strange hospitality to which the guest Smith is indebted is indeed multiple—there are many ways to "put it." As a guest, Smith occupies an ethical position distinct from the son or even the patient. Dependent for his life not on the interiority of a will but on the exteriority of care and a postmodern, machinic kindness of strangers, Smith's relation with the doctor is first and foremost an encounter of kind-ness—qualitative difference. No "expression" or communicative exchange brokers the relation, and even a simulated intersubjectivity can only occur retroactively. Only after a thoroughly contingent—even miraculous—awakening can the care and relation of the doctor and Smith be articulated *as a relation*. Thus orphaned—of "no relation"—Smith is in some sense, as his mother puts it, "lost," outside of the rhetorical regime of the family. Yet this "loss" is hardly lacking—in the meantime, as it were, Smith has plugged into, been connected to, an entire machinic rhizome.

Smith's relation to the machines and care that animate him—perhaps like our relation to cinema—traverses multiple registers, none of which are fundamentally acts of expression. The turning of his body, the connection of a feeding tube, the administration of a drug—these are recipes or procedures concerned not with representation or even communication but with a signifying discipline and desire—the desire to discipline and cultivate Smith's body back into "life." Even after awakening, of course, the discipline continues: therapy and therapists, operations and drugs, foster the possibility of Smith's return to "normality."

Yet this discipline is not confined to technology or even medicine in the usual sense. The shouts of his physical therapist—"Give me one more lap!"—remind us that the ordering of Smith's (and Dockery's) body is also a rhetorical enterprise. Cronenberg's film—perhaps for good reason, as we shall see—offers the possibility of a line of flight from the maternal register of this rhetoric, as the film *engenders*, and does not merely *express*, the syncope of the coma. Even while it attaches the disturbing vitality of Smith's "flashes" to the maternal body through the expressive force of its narrative, a narrative in which the comatose body speaks, the film undoes any attempt to arrest the ambiguity of the coma. Figured by a cut, a fracture "between" scenes that is the very MacGuffin of the film, the coma remains uninterpretable, a blank that is not a lack. More than "offstage" or "behind the scenes," the coma is imaged as an interruption, the very synaptic fissure which makes cinema, subjectivity, possible—the "gap" or interstice between two cites or sights.[38] The film features, of course, the disciplining of Smith—by the media, as well as his "lost" familial order—but this discipline instills yet another affect in the viewer: becoming-comatose. For if the asignifying "interstice" of both the cinematic frame and the coma—in the dead zone these entities are indiscernible—images nothing but time, it thwarts interpretation even as it carries out its effects—an encounter with, but not an expression of, the future. Aha.

At the level of "consciousness," this encounter literally does not occur—it registers a blank. And yet on the register of the film's movement, the very condition of possibility for a narrative that would attach the life of the coma to the maternal body, this blank *is* exuberant production itself, the proliferation of images, the visualization of multiplicity called cinema. And perhaps the most singular figuration of this interruption is the silent "click" with which I began, the hang up on a lost mother on the phone, long distance. This hang up *cleaves*—sticks together and divides—the maternal body and the coma, as the coma *contacts the mother without her knowledge.*

   This notion of "contact" extends to the viewer of *The Dead Zone*. Steven Shaviro, writing of a notion of cinema as "contagion," notes that this shift in the premises of film theory leads away from a fixation on "identification." "When I am caught up in watching a film I do not really 'identify' in a psychoanalytic sense ... It is more the case that I am brought into intimate contact with the images on the screen by a process of *mimesis* or *contagion.*"[39] Shaviro, after Benjamin, characterizes this contact as a "tactile convergence," an encounter between viewer and image that implodes any vision that would render the visual in terms of a distance between a "subject" and an "image." Instead, the viewer enjoys and suffers a strange intimacy with the screen. She is composed as a multiplicity, neither inside nor outside the image, a medium or Möbius body that conducts intensities and flows. Conducting a coma, *The Dead Zone* brings the viewer into contact with the blinding, impossible vision of interruption. It interrupts vision and even consciousness as it induces the syncope of experiencing oneself *being interrupted*, becoming-comatose. Shaviro, via Bataille, describes an "anti-vision" of film that uncannily mimes the very content and experience of *The Dead Zone*.

> We see that which exceeds the possibility of seeing, that which it is intolerable to see. And it occurs in a time of repetition, without a living present, a time that linear narrative cannot fill.[40]

King's novel, while not provoking the interruptions that the rhetorical operations of the film effect, nonetheless allegorizes the encounter between the strange emergence of life called the coma and any attempt to narrate this gift of the future. The very experience of the coma—what, in the film, will have been an interruption, a *caesura*—is narrated through the tropes of childbirth. If Cronenberg's film referenced the maternal body only retroactively, after the return to life, King's novel delivers the coma itself as a labor. It is, however a labor not of the mother, but of the fetus. Narrated in the third person, the comatose character Smith nonetheless adopts the agency that has been bestowed upon the late-twentieth-century fetus. For Johnny in a coma, becoming-lively is becoming-fetus.

> It began to seem that he was not in a hallway at all anymore but in a room— *almost* in a room, separated from it by the thinnest of membranes, a sort of placental sac, like a baby waiting to be born.[41]

A "baby waiting to be born," of course, is not a baby at all—it is a fetus. Yet the becoming-lively of the comatose body shares in the sovereignty of "his majesty," proleptically severing the placental sac that is his livelihood, obliterating the future through his own, impossible, action.[42] Almost in a room, a sleepy hollow between life and death, the sovereign fetus called the comatose body must choose:

> Choose, Something inside him whispered. Choose *or they'll choose for you, they'll rip you out of this place, whatever and wherever it is, like doctors ripping a baby out of its mother's womb by caesarian section.*[43]

This odd persistence of fetal autonomy in the context of the machinic hospitality of the coma testifies to the difficulties this hybrid life form poses for narrative. It also attests to the enormous power of a self-birthing discourse of the fetus, a self-birthing possible only within a historical and rhetorical moment that evaporates maternal bodies. If no mother's body is necessary for a fetus, one is surely not required for an entity that lives off the tropes of the fetus, some machines and some care, "whatever and wherever it is."

And yet the maternal body does not simply disappear in the context of the coma—it is cleaved, severed and attached, at a distance. In her labor, Sarah attaches *her* maternal body to Johnny's coma:

> [A]t some point in her extremity it occurred to her that she was in the same hospital as Johnny, and she called his name over and over again. Afterward she barely remembered this, and certainly never told Walt. She thought she might have dreamed it.[44]

Substituting for a more conventional, accidental, and fantasmatic repetition of an absent lover's name, this iteration of "Johnny" connects labor to the coma, as if Sarah could deliver Johnny out of his silence. And yet for all its repetition—one of the characteristics of technoscientific certainty—this call remains epistemologically uncertain: "She thought she might have dreamed it." The maternity of the coma is thus tangled—both named and, possibly, dreamed, in the placenta and just plain out of it, in "extremity."

So too is Johnny's first experience of his "second sight"—the recognition that his Doctor's mother is, indeed, alive—experienced *as labor,* labor at a distance, virtual labor.

> ...*a baby, ooooh the labor! the labor is terrible and she needs drugs, morphine, this* JOHANNA BORENTZ, *because of the hip, the broken hip, it has mended, it has*

gone to sleep, but now it awakes and begins to scream as her pelvis spreads to let the baby out. one baby. two. and three. and four. they don't come all at once, no—they are a harvest of years, they are.

"The Babies," Johnny lilted, and now he spoke in a woman's voice, not his own voice at all. It was the voice of a woman.[45]

Ventriloquizing maternity, Johnny expresses the pain of birth at a distance. While narrating the events of the birth in the third person, Johnny eventually becomes-woman, connecting the coma victim's becoming-lively rather precisely to the tropes and experiences of birth, births whose futurity—the "Harvest of Years"—are imploded into a continual labor. Greeted with uncertainty—his doctor does not, at first, believe—Johnny cleaves the coma unto the maternal body, joining it to the present and ripping it out of its "place" in a time out of joint, out of pelvis, the dead zone, virtuality.

While this dead zone is imaged as a cut in Cronenberg's film—it is never filled in by narrative, "explained" to the viewer—King's novel declares it to be, in some sense, unthinkable. Subject to a battery of tests, Smith's capacity to summon images—his imagination—is mapped by an EEG. Asked to think of a picnic table on the left side of a lawn, Smith sees a hammock instead. "It's the weirdest thing. I can't quite ... seem to think of it. I mean, I know what it is, but I can't see it in my mind."[46]

This gap or interval in the possibility of imaging is not a loss. Swapping images for effects, Johnny's "dead zones"—imagination's finitude—live at a distance, virtually, perhaps with his Doctor's mother. Not simply a blindness traded for an insight, Smith literally becomes a medium, a site for the collision of virtuality with actuality. This transformation is easiest to articulate in those instances where Johnny accesses the future—unseen becomings are stamped on Johnny's body, as he is seized by, with, the future, virtuality proleptically becoming actuality. Deleuze and Guattari note that this passage from the virtual (the realm of concepts) to the actual (the realm of functions and lived experience) traverses what they call a "dead time."

It is no longer time that exists between two instants; it is the event that is a meanwhile (*un entre-temps*): the meanwhile is not part of the eternal, but neither is it part of time—it belongs to becoming. The meanwhile, the event, is always a dead time; it is there where nothing takes place, an infinite awaiting that is already infinitely past, awaiting and reserve. This dead time does not come after what happens; it coexists with the instant or time of the accident ... [47]

Hence the impossibility of imaging this "dead zone" or "time of the [Johnny's] accident": the virtual, far from being opposed to the actual, constitutes it. But it is of another order, a register accessed only through a discontinuity. To move from the virtual to the actual is to become, a traversal imaged only through the intermezzo of sheer difference, the shattering, continual variation that is the interstice of film, the coma, or, as King implies of Smith's brain "a television. It is on, but not receiving a station."[48] Elsewhere, Deleuze and Guattari map this discontinuity in terms of the connection of (desiring) machine to (desiring) machine:

> Far from being the opposite of continuity, the break or interruption conditions this continuity: it presupposes or defines what it cuts into as an ideal continuity. This is because, as we have seen, every machine is a machine of a machine. The machine produces an interruption of the flow only insofar as it is connected to another machine that supposedly produces this flow.[49]

Rather than positing a dialectic in which each interruption is configured within the same, this diagram of disconnection isolates the mechanism through which maternity is both referenced—virtually—and dislocated—actually—in discursive renderings of the coma. Mobilized in order to narrate the eruption of life out of a machine, maternity becomes iterable, citable, and detached from any specific maternal body. The very disturbance to family the tropes of maternity would suture—life sprouting from technology, getting distributed—enables maternity's detachment from any corporeality whatsoever—the mother remains virtual in the hang-up. This virtual return of maternity—once evaporated by the discourse on fetal rights, now crucial to the narrative of the coma—is actualized not in terms of a multiplicity called "mothering," but as a connector to another machine: family. If the virtual maternal—"Brain-Dead Mother Has Her Baby"—operates at least in part through its capacity for differential repetition, this repetition makes available both a hybridization and a reproduction.

Familial Futures

Networked with other films and novels of the period—such as Michael Crichton's adaptation of the Robin Cook novel, *Coma*—*The Dead Zone*'s investment of the coma victim with the family becomes articulable within the matrix of tensions surrounding the "neomort" or "living cadaver."[50] More than a metaphysical distinction, the emergence of brain death was associated with the increased "procurement" of organs for transplantation. This tension—the uncertainty of determining death in the age of life support and the need for such certainty in an organ donation market—comported

Crichton's *Coma* in the form of organ harvesting—the ascription of brain death to otherwise healthy patients for the purpose of organ acquisition. *The Dead Zone,* rather than manifesting an anxiety of the body's new possibilities as commodity, narrates the relentlessly futural character of corporeality, a futurity most easily accessed through the donation of the "living body" that houses the "dead" brain to others, a future continually referenced in materials that recruit organ donors.[51]

    This future—another body that might live through organs of the deceased—has a particularly familial flavor, as the family is both a locus of the decision to donate organs and, sometimes, their recipient. When Monte Burns, the brutal but affectless capitalist on *The Simpsons,* learned he had a son, he failed in his attempt to simulate the love of a parental bond but remarked, with great sentimentality, "It's good to know... that there's another kidney out there for me."[52]

    Perhaps the location of this sample—a flickering animated image transduced to video—evinces the character of this new form of life maintenance. Living tissues no longer reside within the confines of an allegedly autonomous body, but are instead contingently networked with differential futures, futures of animation. New segments of life are being produced through the organization of donation and the dosing of immune systems. This production of life—"The Gift of Life"—emerges only through the transforming of selves—futurity emerges for the transplant recipient only through the disciplining of the very contours of the self, a self sculpted both of drugs and discourse, a rhetorical and pharmaceutical hybrid.[53]

    But this futural character of contemporary corporeality—its ability to act in a future, without consciousness—poses severe challenges to any system that would procure and allocate human organs. If such dispensations are to emerge from a citizen, such subjects must enjoy the proleptic ability to act in the future, after one's death. In other words, organs must become concepts—treatable and articulable in terms of their ability to become rather than maintain.[54] They are treated in terms of their capacity for difference.

    This extension of agency past the life of the subject paradoxically depends upon the agency of the family. For if the body is to be a transaction site in the absence of a subject, it is the family that brokers, executes, this transaction. The very death of the subject that enables the gift of organ donation thwarts any testimony of the gift's propriety. Thus the family—and not merely writing, a contract, a donor card—is called forth to ventriloquize the brain-dead subject: "To be an Organ & Tissue donor, even if you've signed something, you must tell your family now, so they can carry out your wishes later."[55]

This testimony to the future perfect desires of the comatose—
"This is what she would have wanted"—is not, of course, without its share of ambiguity, an incapacity which continually haunts and perhaps thwarts organ donation. In the context of the termination of life support—the proverbial "pulling of the plug"—this ambiguity has lead to the invention of a new speech act in the United States, Australia, and Europe: the living will.

Thus while it is clear that the technologies of life support and the legal armature of brain death mark a decisive diagram of subjectivity—even in brain death, a subject persists, variously entangled with both the law and machines—comas also mark new configurations of power that emerge out of the "family." Even as the technologies of life support and organ transplantation—the twin vectors that compose the coma as such a site of anxiety—make possible new modes of human hybridization, it is the discourse of the family that ensures the continual attachment of a point of view, or "center," to the distributed rhizome of growth that embeds comatose bodies. Indeed, the value of the family almost seems to emerge out of its ability to govern the proliferating disturbances that such technologies pose for "life."[56]

This position of families as the "governing" body for coma patients is outlined in a recent court decision in Pennsylvania, where the right to terminate life support from a patient in a permanent vegetative state (coma) is granted to families, even in the absence of a living will or "advance health care declaration." Note that this remarkable power of interruption is withheld from women themselves, as an advance health care declaration becomes null in the event that a woman is suspected of pregnancy.

> A declaration by a pregnant woman will not become effective unless a physician has determined that the life-sustaining treatments either (a) will not permit the live birth of the unborn child, (b) will be physically harmful to the pregnant woman, or (c) would cause pain to the pregnant woman.[57]

Here the agency of women's bodies is both referenced and obliterated. Literally cleft from her own "living will," the proleptic desire of woman becomes null and void, or at least without effects, not "effective." Under this rubric, the agency of (possibly) premenopausal women becomes haunted by the specter of pregnancy.

At the same time, this cleavage of the living will from Pennsylvanian women reinscribes the limbo agency of the maternal body itself. Capable of birth, capable of pain, the maternal body is not simply erased in the desire to foster the fetus. Instead, it is cleaved, detached from the past desire of a woman and attached to the future capacities of the maternal body, capacities that, by definition, are virtual—attached to the future as much as a mother and life support—and not actual. Thus it is the virtual maternal that is detached from the body of a desiring woman, a virtuality that enables the literal propagation of a family even as this propagation is detached from the specificity of any individual female subject. Somebody in Pennsylvania read "Brain-Dead Mother Has Her Baby."

The emergence of a body that uncannily lives up to Heidegger's dictum concerning the "essence" of technology—the coma is Nothing, technological—would seem to remind us even more forcefully of Spinoza's refrain: We don't know what a body can do.[58] Endowed with new capacities of distribution, contemporary bodies cultivate futures through routes other than reproduction. Connected with life by machines as well as so-called biological multiplicities, human and otherwise, the networked body of the coma relates not to a future of knowledge, but of contingency. A virtual corporeality, the comatose body is, in the phrasing of philosopher Elizabeth Grosz, "befallen by the future."[59] Its genealogy is, then, uncertain, available for difference, aleatory to the assemblage of family that would send a name, and the flesh to bear it, into the future. The comatose body encounters a future that cannot be *divined*—no reliable algorithm exists to signify its future, to predict with precision the actualization of its virtuality.

> He tried to protest, to tell her that he didn't want to do great works, heal, or speak in tongues, to divine the future ... He tried to tell her, but his tongue wouldn't obey his brain.[60]

In the silence of the coma, betwixt tongue and brain, a dead zone, the American family speaks. Despite the much-noted "decline" of the family and the threats that various postindustrial configurations allegedly pose to it, the family ultimately governs the comportment and treatment of the comatose body. If the comatose body troubles our understandings of the medical and legal spaces of life and death, it is through the discourse of the family that this odd, resistant body becomes governable, even divined.

"Here's Johnny!"; or, Textual Delivery

At the Westfall Health Care Center in Brighton, New York, a twenty-nine-year-old woman, comatose for ten years in the wake of a car accident, developed a swelling of the belly, indicating a pregnancy. One media account narrated the 1996 event in terms of a textual materialization or delivery: "It is as if a particularly diabolical textbook problem for budding ethicists has been transposed to grim reality in upstate New York."[61] What was remarkable about this horrifying rape was its immediate management by the discourse of familial reproduction, as the woman's family determined that she would want to bear the child.

> The New York woman had been a devout Roman Catholic. Says Ellen Moskowitz, a lawyer and ethicist at the Hastings Center, "It seems reasonable to conclude this is the kind of decision she would have wanted."[62]

By continuously invoking the values of the famed yet anonymous woman in terms of the discourse of pro-life, the family was able to render a paradoxical decision "against abortion"—paradoxical because the condition of possibility of abortion was itself life support, itself the result of another incalculable decision. Carrying the child to term—with the help of life support—was continually warranted through recourse to the woman's pro life values, without which such a birth out of life maintenance would have been, in Hartouni's phrasing, "virtually unintelligible."

And yet perhaps it is only virtually that this birth is made possible. The delivery is entangled with the enunciation of the woman's family—it is literally enabled by a statement, along with some other machines. The statement or command—"Do not abort!"—is utterable, in this instance, only to the extent that it can articulate the alleged desires of the woman. These desires are *neither here nor there*, embedded in the discourses and practices of the past and yet strangely effective on the present and the future. The family functions as a relay—or ventriloquism—of desire.

As desires, these discourses and practices do not refer—they are promises of a capacity to be affected and perhaps transformed. "I want" is less a reference to a lack than a sensitivity to an other. So too is this sensitivity itself characterized by its citationality, a capacity for the repetition and drift of geography and time. It is in this sense that the discourse of the woman in question is virtual. Indeed, it is the actualization of the woman's desires *as virtual* that enables this birth, as her desires are treated as detachable from any particular actualization, "what she would have wanted."

And how are these virtuals actualized? The instantiation of the woman's desires takes place through both the enunciation and narratives of family. The woman's ghost is summoned to speak, a living ghost. This ghost is provocative of, but does not represent, the past. An uncanny stranger hailed as much by the citable character of the familial narrative as it is by the life-support mechanisms that enable its development, the baby sprouts out of familial discourse, a discursive budding or hybrid that allows for the reproduction of family "value" into the future, a fleshly delivery that emerges partially from an utterance.

Crucial to the decision to enable this birth, the first live birth from a continually comatose body, was the ascription of agency to the mother even as it was erased. Doctors noted that

> the natural forces of labor could possibly be intact in the woman, but without voluntary cooperation in pushing, the delivery would require forceps or a vacuum.[63]

As with our earlier phone call, the connection to the mother is established, but without her knowledge. Familial production operates here not as the reproduction of individuals but as the proliferation of the familial itself, the actualization of virtual entities or values bound not to consciousness but to the production of other live bodies, live bodies that emerge out of the rhetorical, virtual space of "decision." These remarkable pragmatics of speech emerge out of a virtual, familial response to that old, supposedly enigmatic psychoanalytic puzzler: What does woman want?

Ungovernable by the taxonomies or articulations of medicine, the comatose body renders not merely another human, but an occasion for the exercise of familial power. The nonstop public performance we call family, rather than wounded by the strife of the uncanny space between life and death, is constituted by it. The family speaks, it decides, and children are born. With this child—a two-pound, eleven-ounce boy born March 18, 1996, let us call him "Johnny"—a new speech act is born, a discursive birth that performatively constitutes the flesh of the future: "We are Family."

E   I   G   H   T

"Take My Bone Marrow, Please":
The Community in Which
We Have Organs in Common

Recipe: Segment must be accompanied by large blue and white Igloo cooler with mysterious contents. Options include: Organ Piñata, a large plaster of paris kidney, painted pink with red swirls. Spray paint will do. Piñata is stuffed with tiny candy bars wrapped in organ "donor" cards that sign rights for future sale of organs over to Wetwares Inc. Organ Rover: Igloo cooler mounted on toy monster truck, pursuing improbable trajectories linked by a global positioning system and Dr. Jack Kervorkian's pager.

THIS SEGMENT will emerge out of a problem, the problem of how to give something that you quite simply don't have. Perhaps only a scholastic—one who owns a franchise on the production of distinctions—would be troubled by this problem. "How many absences can you fit on the head of a pin with angels dancing on it?" they might puzzle. "How could you possibly balance anything on a pin that wasn't there?"

Nonetheless, you see the problem. From a certain conceptual perspective, you just can't do it. Giving, for example, can only emerge from the given, that which is already there. You know the story. It's tired. You go to give something, like a giving of yourself, and it's just not there. So to give this story some spin, I'll make it *give*, to see what the giving entails.

An Initial Public Offering

I am thinking of starting a company. I often do this in between moments of thinking up screenplay ideas that I will work out later. I take sociologist of science Bruno Latour at his word, that to think technoscience is an expensive game. The skeptic—although why should it only be the skeptic who would be interested in such an enterprise, how about a hobbyist?—has to amass enough capital to test out specific claims of a technoscience, such as the effectiveness of an immunosuppressor like Cyclosporine or the veracity of a genetic map of *Cannabis indica*. In short, the skeptic must build a lab.

I love this vision of a democratic science—let a hundred labs bloom. I see neighborhood labs competing for space with ping-pong tables in the few community centers and organizations we have left. I have a vision of DNA sequencers crammed in church basements, trailing-edge technology that allows Girl Scouts to test out theories of alien abduction, producing counternarratives about the genetics of cancer. In place of conspiracy theory we will have a multiplicity of technoscience.

Of course, the whole reason these sequencers are on the junk heap is that they are trailing edge. Sure, we could produce counternarratives, but the truth, ah the truth would probably still be "out there," somewhere else where more money and information clusters. So maybe, just maybe, we need to line up some capital with these packs of hobbyists, Water Buffalo lodge members, bridge clubs, women's shelters, mummers, yes, even Masons, and hack technoscience. I have even approached ISIS, that gang of counternarrative machines out of Amherst that cleans up toxins, studies multiple chemical sensitives, the future of cryptography, privacy, becoming, and they said yeah, we could use the dough, too.

So here is the plan for my new business: I want to start the first futures market in human organs. Get paid now for a future interest in your possibly healthy organs. Only to be harvested after death, of course. So far, I have some decent ideas about how to get such a system up and running—this idea is hardly my invention—but I am going to need a lot of help. I need you to help me. In helping me or resisting this scheme, think about the difference a distributed technoscience would make, one in which the means of knowledge production were as deterritorialized as capital itself, in which the means to produce or resist narratives of technoscience moved across the globe as quickly as cubic zirconium on the Home Shopping Network, where technoscience "kiosks" had more than Web access but entire labs, and the signs out front of proliferating franchises boasted, speaking of their narratives, claims, and outrageous arguments they have produced, "Billions and Billions Told..."

Of course the problem is that if this deterritorialization of techno-science—what Deleuze and Guattari have characterized as a nomad science[1]—is to take place, bodies too must be deterritorialized. Not that much flesh hasn't passed through the ultracentrifuges of late, late, late, late capitalism already. But a short-term supply-side shock has overtaken organ donation: as technologies of organ transplantation have improved drastically, so too has demand for human organs increased, even as donation rates have been relatively flat.

But if the deterritorialization of organs allegedly proper to the body has been snagged by the identity practices of the present, this tangle is involved with the piety and privilege of "life." Organ donation discourse is bogged down by the sappy, oedipal recuperation of organ donation as the "gift of life." In the Transplant Games—that sporting event where one is actually *required* to take drugs—transplant recipients receive their medals from donor families—family members of those who have given their organs in death such that others might vault, sprint, swim, and live. This tactic of oedipalizing the organ—plugging it into a system of reproduction, the giving of an organ is the giving of "life"—seems to have reached its limit, though. The economy of organ transplantation is an economy of scarcity—thousands die each year waiting for someone else to die so they won't. So before the American postoedipal body can be turned into a futures market for a distributed technoscience, new tactics must be invented to deterritorialize bodies coded by autonomy. This segment is an early effort to effect this deterritorialization. It is my shtick for (ad)venture capital.

I begin cautiously, with bone marrow, a human tissue that is not—like other human tissues such as kidneys and livers—in an economy of scarcity. This is so for a very specific reason—the giving of bone marrow is not a zero-sum game. The human body is an exuberant bone marrow production. So I begin by trying a new tactic in the liberation of marrow, a tactic that draws not on some notion of a gift of life that sutures finitude in the face of that thing we all allegedly have in common—death—but one that draws on the laughter and difference that attends to the experience of the permeability of the body, its character as a multiplicity entangled with other multiplicities.

The Community in Which We Have Organs In Common

You know, where I live, things have gotten out of hand. Take my bone marrow. Please. No really, I mean it. Take it. Take it now. I mean, I have a cooler for you right here, you've seen the footage on TV, technicians running—very carefully—with what looks like a lunchbox, mist streaming off the lid. Not really something you'd want to eat, but hey maybe you can get together with it another way. You know.

I wish you would take my bone marrow. I mean really. It's not such a big deal—I am all the bone marrows in history. Take all you want, we'll make more. My marrow and your marrow, they get together. See what I mean? In the meantime—To-marrow, even—more bone marrow. For me!

So take it. What are you waiting for? I know, there's just no money in it. You just don't hear rumors about guys waking up naked after a night on the town, sitting in a tub full of ice, with a hangover and no bone marrow. No, you just don't hear people worrying about Mickey Mantle leapfrogging some poor sick fuck from North Dakota for a bone marrow. Bone marrow don't get no respect, you ask me.

But hey, take it anyway. You never know.

So you probably, frankly, don't want it. I can't give this shit away sometimes. But still take it. Cause if you don't take it, what the fuck am I gonna do with it? Nothing. Me—I got plenty. And there's more where that comes from, I mean we can make arrangements. You know what I mean.

So you're thinking, okay, maybe I just will. Don't mind if I do. But wait just a minute. Take just one minute. I said to take it, yeah, but that was then. And this is now, old marrow stealer. And I say you can't have it.

Just kidding. Calm down, it was a joke. I wouldn't change my mind, not like that. You can have it. I said you can have it.

But can you really take it? That's what I think, I think you can't fucking take it. No way. Can't take it. That's what I hear about you.

"Says who?" says you. I says so. "I" says you can't take it. "Me," that's who. Mr. Subject of the sentence over here. I says you can't take it. Never.

Take it anyway, would you? Just take my bone marrow, please!

No, you just can't take it. I'd have to give it to you now, wouldn't I? I would have to say—as I keep reminding you—"Go ahead. Take my bone marrow. Please."

Oh, no, not gonna take that from me, you think to yourself. I don't take that from nobody. Rather—since you never know when a grammar cop is around—I'll take it from nobody. I've read up on this stuff. I know about the Merchant of Venice, the pound of flesh. One pound of meat and a heap of trouble. No way. Don't give me any of that.

Just take it, ok? I was polite enough. I said "please." I don't want to give you any trouble. Just the bone marrow, and nothing but. All you need is a large needle—is there a doctor in the house? There's no need take anything from me. You couldn't if you tried. But I'll give it to you.

"You" still don't get it. No, you just don't get it. You "just" don't. What you get is a bone marrow, sure. But there are other considerations. Like maybe "you" aren't the one that gets it. "You" aren't one of the ones that gets it ever. No "one" gets it. But sure, I'm giving out more than I am taking in. So some body must get it.

Nobody.

Well, you're thinking, if I can't get it, then nobody should. It's only fair that nobody should get what I don't.

So good. We understand each other at last. Nobody gets my bone marrow. So take it. Please.

I mean somebody might as well. If nobody wants it, you might as well take it. I know. You're doing a double take now. Thinking I might be on the take or something. Like now he wants to give me some marrow. But what about before?

I said, "Take my bone marrow. Please." Really I did. But you couldn't take it. Now that I said nobody could have it, I thought—for surely nobody could take it—I might as well give it. But not to "you." Oh, if we're a match, we'll take it and put it to good use. Or we'll give it to somebody else. But not "you."

I know, I know, cheer up. "You" never get anything. "I" know how it is. But the way this whole bone marrow thing works—and it works, really it does—is that "you" can't take it. I can give it. Or can "I"? How can "I" give what nobody took?

Yeah, yeah, the needle. But I'm not gonna give myself a needle. No way. That's somebody else's job. Not mine. Which means, within our little give and take, that somebody's going to give me a needle. A big, fat one.

Shit. I really can't take needles.

Oh, I've given blood, yeah, had plenty of allergy shots in my youth, a huge shot of painkiller when I broke my leg at age two. And I dig—really—those moments in junky films—like *Drug Store Cowboy* or *Bad Lieutenant*—when the camera highlights the needle's entry into an arm. I mean, how long did that scene in *Bad Lieutenant* last? It was the sex scene of the movie. Talk about pleasing.

Obviously, you're saying.

But I can't take needles. So somebody is going to have to give me a shot, a big one. And before that, there will be all this paperwork to be done. Have I been to a malarial zone lately?

Have I?

So along with a needle, somebody's gonna have to put me under a bit of surveillance, give me a looking over. Which I can take. Really. No problem.

Test me for HIV? You wanna know where I've been these past few years? Childhood diseases? Already told you about the broken leg, but that doesn't matter. When do I get to tell you about my allergies? And I'm a vegetarian, you know.

So take my bone marrow. Please. I feel like maybe we are becoming friends. Take my marrow—if you take it, if I give it, we'll have that going for us. And it'll have nothing to do with "you" or "me." Cause it was never mine in the first place. Please?

N   I   N   E

Wetwares; or,

Cutting Up a Few Aliens

wetware /wet'weir/ n. pl.

[prob. from the novels of Rudy Rucker] 1. The human nervous system, as opposed to computer hardware or software. "Wetware has 7 plus or minus 2 temporary registers." 2. Human beings (programmers, operators, administrators) attached to a computer system, as opposed to the system's hardware or software. 3. 'Taint software, 'taint hardware (Doyle, 2002). See *liveware, meatware*.

> *The Online Hacker Jargon File,*
> modified by Richard Doyle

(1.0)

EVERY FEW months, my mailbox houses an improbable visitor. *Transplant Video Journal*, a simulated television news program produced by Novartis Pharmaceuticals, finds itself sandwiched among surreal garden catalogs, earnest supplications from vaguely left nonprofit organizations, bills, and postcards from friends working hard to keep in touch.[1] The sight of the tape in my mailbox, wrapped in the proverbial plain brown wrapper, its return address cryptic, excites. My father, after all, was a mail order pioneer.

(2.0)

The small bowel is not the most transportable configuration of human tissue. More an ecology than a well-defined, autonomous organ, the small bowel teems with a zoology of fauna that carry out the intricate and thankless tasks of human digestion. Beyond the mechanics of peristalsis and the transformative action of saliva and stomach acids, human digestion is outsourced to various stains of bacteria.

Thus, unlike many of its corporeal neighbors—kidney, liver, heart—the small bowel is transplanted only with much difficulty. According to United Network for Organ Sharing, fewer than three hundred small bowel transplants have ever been performed. A xenograft, small bowel transplants raise the question of the identity of the *donor:* an entire inhuman ecology is being deterritorialized.

(3.0)

I have been working all day in my office, my usual daytime habitat. Sometimes I feel like a cop, surrounded by stale coffee, continually eating sandwiches at my desk, interrupting each new task with a more recent "emergency." Usually, though, I just stare at a continually changing screen for most of the day, pausing to argue, joke, and respond with my students and colleagues in the classroom, hallway, or office. My eyes burn from the optical gymnastics of font and pixel. Bausch and Lomb Computer Eye Drops are the ticket for this media burn. They coat, soothe, and protect my eyes with a formula designed for the stress of today's optically saturated work place.

(4.0)

To "deterritorialize" something, the way some authors tell the story, is to connect an entity to elsewhere. A hand deterritorializes a tree branch when it becomes a tool for digging. Disconnected from a tree, its end becomes pointed, sharpened in an encounter with the earth it would transform; forest becomes farmland connected to maize, becomes pasture, burgers grown in the former rainforest are routed via drive-thru windows into waiting stomachs, a small bowel is transplanted. Some organisms deterritorialize through reproduction—a maple seed helicopters elsewhere, giving the lie to the notion that plants are motionless. Others also induce movement through rhetorical transactions: bee semiotics—an iterable dance—establish a flow of nectar to the territorialized hive, the promise of payment vectors spices to Northern Europe from Indonesia. Money goes from one account to another. Sometimes, it is in more than one place at once, *particularly if one of the places is the Grand Caymans.*

(5.0)

The screen is an immaculate blue, an image of waiting. I once toured the CNN news studio, where I became fascinated by the blue felt background that the meteorologist gestured toward as she indicated a virtual low pressure system here and an invisible radar photo there. I, of course, saw nothing but blue. On a blue background, everything seems promising, an icon of the virtual to come. A button on the remote is pushed and the blue is overtaken by black.

(6.0)

In recent years, computers have been radically deterritorialized as they became hooked up to networks and webs. Personal computers emerged from their basements and closets and became telecommunicative devices enabled by the enormous growth of the Internet. The World Wide Web—a hypertext system designed for organizing research and communication at a high-energy physics lab—was itself deterritorialized from its Swiss milieu, sprouting into the entertainment-cum-commerce-cum-research tool it is. Emphasis on the cum.

(7.0)

As soon as the image arrives, it cuts itself open, slowly peeling back a layer above the "first" image. The cut—in the middle of the frame, backgrounded by fuzzy biomedical scenes—opens into an opened body, a gloved hand probing and stroking layers of opened abdominal flesh. "We'll go into the operating room for a small bowel transplant," Wayne Armstrong, our host, tells us. I wonder just who, or what, is receiving the transplant.

(8.0)

It is by now a commonplace that this deterritorialization—the growth of computer networks as acentered, distributed webs, networks that, once connected, aren't "anywhere" in particular—emerged, in part, out of the strategic anticipations of defense intellectuals. DARPANET—the primordial growth of connectivity that would become the Internet—was built to distribute the knowledge base that made possible the design and testing of nuclear weapons. In this scenario, a network of computers would enable weapons design to proceed even in the face of a strike on any one of the computers. Networks emerged as proleptic adaptations to the continual specter of nuclear destruction. *They did so by becoming redundant.*

During this frenzied connection of computer to computer, some writers have speedily rendered this new distributed technology in terms of its impact on our selves. "Like a chip factory running flat out, the American self plops off the conveyor-belt with wired flesh."[2] The appropriately named Krokers plot the technobirthing of this American self according to a global software update called late capitalism: "The American Algorithm? That's America as the Operating System for global culture at the beginning of the 3rd millenium."[3] Here the earth itself, when it's not busy getting in drag and simulating a superorganism so that it might pull off the Gaia routine and fool evolutionary biologists, atmospheric scientists, and new age ecofeminists, logs on to every cognitive dissonance and alimentary disturbance of the American psyche. The surface of the earth—the skin disease Nietzsche's Zarathustra characterized as "man"[4]—swarms with the most banal episodes of human consciousness: "writing software is actually like writing out in code what it means to be an American."[5] This cognitive excretion—what is inside, comes out—downloads the American subjectivity franchise into a global flow. Episodes of *Melrose Place* are folded into Java applets, secret formulae for McDonald's special sauce find their way into an algorithmic handshake among modems, plans for MX Missile deployment get pasted into a subroutine that renders menu options in Windows 95, the notional missile defense project is launched.[6] The virtuality that is the American psyche, the Krokers note with horror, implodes into a nauseating collision with an entire Cosmos: Earth become Disney. Planet Hollywood. Universities.

(9.0)

There is a soil fungus that, properly treated, allows humans to tolerate the flesh of others. It enables a graft of an alien organ, donated or purchased flesh. Cyclosporine nurtures the great network of organ procurement and transplantation that now generates thousands of transplantations in the United States alone. These newly transportable human organs are grown, then, in a medium of Cyclosporine. They emerge out of an ecology of fungus.

Fungi are themselves strange synecdoches for a network; spores emerge out of the fruiting mycelium, a massively connective network of fuzz often grown in your refrigerator—old tomato paste, a tired piece of cheese, leftovers. Spores deterritorialize the mycelium, propagating it through time. The dense spatial connectivity of the mycelium is compressed and launched through, into, the future. Time travel . . .

(10.0)

Writer William S. Burroughs located a remarkable capacity of texts to encounter the future—their capacity to be grafted or cited into another context, text, conversation, or dream. Burroughs was repulsed by conventional concepts of immortality—what he called "vampire schemes"—as the desire for stasis, me me me me me me me. Burroughs saw both heterosexual reproduction and Christianity as viral ecologies of the same, the worst habits of a human species at an impasse. Burroughs favored instead the hybridity of alliances with familiars both technological and biological. Burroughs sought to accentuate and cultivate the strange facility of texts and selves to mean and even do something other.[7]

(11.0)

Part of the terror of the Krokers' cyberhorror Theme Park Earth is its irreversibility: "You can check out anytime you like, but you can never leave." The virtual becomes ubiquitous, a roach motel of simulacra from which it is impossible to exit. Inside, we suffer a nostalgia for feeling, a longing for the natural. Unlike the America of old, there's no loving or leaving it. The transformation of America from a nation to a set of algorithms—"The Global Operating System"—is, for the Krokers, a revolution of artifice, a negation of corporeality in a decade of cannibalism: "the flesh-eating 90s."

But perhaps the Krokers' worry over selves is itself in need of an update. It's coded for a capitalism based on repetitive production of goods and services, the industrial moment of scarcity and demand. The "conveyor-belt" motif through which the Krokers render the contemporary diagnosis of self belongs to a moment of linear production, homogenous quality, time rendered into quantity. A self that punches the clock arrives, disciplined, to enact one action after another, a temporal series of events—being silent, disciplined, and, above all, in motion—and takes breaks, loses limbs, periodically rests, takes vacations. The jitself—just in time—is not delivered on anything as sequential as a conveyor "belt": It is catastrophic, suddenly emerging between the nodes of a network.

(12.0)

The peeling image on the screen provokes something like vertigo. From here on, it's image all the way down. The installation of the cut layers the image, revealing not an image below the other but a grafting of one onto another, an accretion. The cut into the video image connects only to more layers, stratified flesh, from outside to inside. The image is exposed in its grafting to flesh; it becomes an algorithm, a recipe for the very connection of image and body: Cut!

(13.0)

Pools and canals reflected excrement mixed with flowers—story of absent legs up the tarnished mirror—In a rusty privy youth bends over—faint drums of memory on his back—green boy of broken glaciers and skin instructions...

> William S. Burroughs
> *The Ticket that Exploded*

Still, if the Krokers' map of "wired flesh" is keyed to the legend of homogenous quantitative time installed with the industrial moment, their understanding of flesh as a cluster of algorithms resonates with the encounters this infomercial is attempting to narrate and enact. *Wetwares* titles an encounter with flesh as a refrain, a repetition of algorithms or recipes of sufficient complexity that only through instantiation can they be experienced. Corporeality names an experience of materiality that is literally inconceivable; only by doing it, as the shoe company named for a missile named for a Greek goddess might put it, can it be done. Mathematician Rudy Rucker names those patterns "inconceivable" entities that are too complex to reproduce in detail.[8] Wetwares are inconceivable not because they sublimely exceed any reduction or representation but because they quite simply cannot be modeled in advance. They cannot, like a complex algorithm, be separated from their enactment. Less lines than flows, wetwares are continually encountering drifts and turbulence as they slosh through time. Less iterations than itinerations, becomings, only as transformations shall ye know them.

(14.0)

For Burroughs, immortality—a grafting to the future—was to be found not in time but in space, through the shattering of a self into becomings.

> We made thousands of cut-ups. When you cut and arrange words on a page, new words emerge...The word "drafted," as into the army, moved into a context of blueprints or contracts, gives an altered meaning. New words and altered meanings are implicit in the process of cutting up, and could have been anticipated. Other results were not expected. When you experiment with cut-ups over a period of time, some of the cut and rearranged texts seem to refer to future events.[9]

(15.0)

Imagine not being able to eat. Not just for a few hours or days, but for years. People in need of small bowel transplants have one thing in common: They can no longer use their own intestine and rely totally on intravenous feeding for survival.

Wayne Armstrong,
host of *Video Transplant Journal*

I find it impossible to obey the narrator. I am simply unable to "imagine" such an oral blockade. The ecology of my mouth connects to the affect of hunger, my lungs with the sting and dose of smoke. I have been on IVs and lost heroin addicts to needles, but as a relation of my body to the outside, hunger is territorialized onto my mouth and stomach.

On the other hand, the imperative "imagine" does make me panic.

(16.0)

The late writer William S. Burroughs rendered both the narrative of and the experience with corporeality as a set of becomings. The Earth, in Burroughs's view, had become a shithouse, an endless swirl of commodity junkies addicted to buying and selling the same old shit, shutting down the "soft machines" by transforming them into need franchises. Junkies were only the most crystalline—and perhaps the most ethical—species of human overtaken by habit, repetition, and the worship of lack. In the face of the beautiful, exhausted human repetitions of violence and destruction, Burroughs welcomed an eruption of the alien and the inhuman that would disturb the very borders of experience. His own method of the cut-up was a disciplining of authorship that might allow what he characterized as the "Third Mind" to be encountered through that most humanist of technologies: the book. The introduction of chance operations into the writing of such novels as *The Soft Machine, The Ticket that Exploded,* and *Nova Express* marked an attempt to articulate human experience not in terms of its repetitions and its closures but of its continual variations, its contingent, complicit itinerations. The very borders of the human body—and those of the book—were mapped in terms of an oozing fluidity: "Shlup Shlup Shlup."[10] Burroughs's work was thus continually and differentially engaged with the interruption of human experience, an introduction to posthumanity. Filled with "blasts of silence," his texts cultivate a reader who can shut up long enough to mutate. In the silence, Burroughs offered recipes for becoming:[11]

> *Naked Lunch* is a blueprint, a How to Book.... Black insect lusts open into vast, otherplanet landscapes.... How-To extend levels of experience by opening the door at the end of a long hall.... Doors that open only in *Silence*.... *Naked Lunch* demands Silence from The Reader. Otherwise he is taking his own pulse...[12]

Swapping redemption for becoming, only mutation can save us now.

(17.0)

I am waiting for the money shot. The image of the stroked layered flesh of the abdominal wall is itself interrupted by an edit: A man walks slowly across an airport tarmac with a blue igloo cooler. According to an anthropologist I spoke with at a conference, some vegetarians feel odd about receiving donor organs. Some meat eaters also feel a cannibalist repulsion to the incorporation of another's flesh. All the more odd that organ procurement agencies should carry these "gifts of life" in a lunch box. Who or what is being fed? Since when was a human organ part of a balanced diet?

(18.0)

Burroughs's composition of an algorithmic or how-to art renders bodies themselves as clusters of memory and instruction—"faint drums of memory on his back—green boy of broken glaciers and skin instructions..."—and Burroughs's practice emerges contiguously to the molecular biological understandings of living systems as informatic economies. As in Burroughs's work, organisms, in this view, were articulated not as sites of autonomous growth or development, but as *topoi* of informatic exchange. The teleonymic characterization of organisms—their only purpose is replication of their replicative cores, DNA—resulted, in its most extreme form, in an oddly Burroughsian picture of "soft machines" inhabited by viral entities with their own parasitic agendas. Biologist Richard Dawkins, for example, recasts humans as lumbering robots for the deterritorialization and propagation of DNA. Here flesh emerges only as the contingent fold of a two-dimensional code into three-dimensional proteins. Protein folding thus occupies the limit position of contemporary life science, a node in an informatic network that is entangled with contingency. It is possible that this fold is in fact composed of contingency.[13]

(19.0)

What shape is the small bowel? A tangle, it somehow is lifted by a pair of human hands into the opened pocket of flesh. Stitched and eventually grafted to the anus, it allows the recipient to eat for the first time in years.[14] Nonetheless, this is no cut and paste operation—the future is nothing if not at stake. The recipient's body, dosed on Cyclosporine, could reject the alien flesh and its bacterial inhabitants. The transplanted bowel is thus stitched to the surface of the skin, rendering its acceptance or rejection visible to the eye of a health care provider at someone's HMO.

(20.0)

Both Burroughs's understanding of language and the molecular biological conception of organisms sprout in the substrate of "information." Here bodies and statements both become occasions for the exercise of coding and decoding technologies and disciplines. Crime, for example, becomes tied to statements not so much of truth about the past than about the availability of codes in the present. A trace of DNA becomes a speech act, a highly regulated testimony to an absent but promising presence: a criminal body. A wire is worn, a body becomes a transducer for testimony, a sex act becomes a wired word. Such an informatic economy does not represent bodies so much as it orders them, organizing them in space and time. "You were there." The impact of a body resides in its capacity to alter configurations of information: Lights. DNA. Action.

(21.0)

One of the most popular explanations for cattle mutilations was alien organ harvesting. In September 1979, "reports of strange bloodless animal mutilations began to appear in the United States, Australia, and Canadian newspapers—accounts of a steer's ear, eye or tongue taken, its genital and rectal tissue excised with cookie-cutter-clean precision, its jaw stripped of flesh."[15]

(22.0)

"The smallest interval is always diabolical."

(23.0)

The recipient's body is prepared at about the same time as the donor's. What is the recipe for the preparation? Once upon a time, there may have been a discussion among family members about funeral preparation. Maybe nothing at all was said. Maybe a driver's license names the deceased as a donor. Perhaps an organ procurement organization is contacted when brain death is declared. The family is approached by an organ procurement professional or a health care provider. A decision is made. Harvesting begins just in time, as soon after the declaration of brain death as possible. The question of donation is "decoupled" from the declaration of brain death, so there is an interval of shock and grief.

(24.0)

This informatic distribution of selves indexes a shift not merely in the self's location but in its possible topologies as well. Being beside oneself entails a set of practices that fold and knot the forms of subjectivity that sprout in contemporary ecologies of capital and media. "I'm not a doctor, but I play one on TV" becomes the mantra for a self that seeks to articulate its various simulacra with a geography and an ontology— on TV, not on TV—capable of locating or finding subjectivity. *This is your brain. This is your brain on drugs.*

(25.0)

This knitting together of various selves of different locales and moments is the very function of contemporary subjectivity, the one still hilariously and hysterically capable of saying "I," of promising to promise, of writing and signing a living will or donating organs. Here it is: "Turn off my life support. No heroic measures." These proxy selves carry out and even endure the work of telling the story of contemporary subjectivity distribution.

(26.0)

I am waiting for the money shot. The image of a stroked layer flesh of the abdominal wall is itself interrupted by an edit: A man walks slowly across an airport tarmac with a blue igloo cooler. "Imagine not being able to eat. Ever." The money shot. The image of the need. According to an anthropologist I know, some vegetarians feel odd about receiving donor organs. All the more odd that they should carry these "gifts of life" in a lunch box. Some also feel a cannibalist repulsion to the incorporation of the flesh of the other organ procurement agencies nonprofit organizations, bills working hard to keep in touch with my slack self feel a cannibalist repulsion to another...

T   E   N

Sympathy for the Alien:
Informatic Ecologies and
the Proliferation of Abduction

Reference Abducted; or, Saucerian Section

ON ANY ordinary day—to the extent that there are such things—I see an awful lot of aliens. Most of them are Grays, with olivine eyes that take up most of their egg-shaped faces. Come to think of it, most of them are nothing *but* face, eyes adorning backpacks, skateboards, bumper stickers, tattoos. And none of them move. Not even a little. They are, strictly speaking, liquid, conforming to the space of their containers, miming the rhythm of their vehicles. When I am not looking, they seem to replicate. I mean, how else did so many of them suddenly appear? There are even several of them on my computer as I write this. "Perhaps he is collaborating in the writing of this right now."[1]

Unless you have been in a coma since 1947 (and perhaps even if you have), you have seen them too. Don't pretend. I know, it's a bit embarrassing. I can't remember the first time I saw one, if there was a first time. It's as if they have always been there, and I simply have no memory of them. Do they, like any fit object of a paranoid narrative, remove evidence of their past existence, wipe our memories clean even as we gaze upon them? Why can't I remember the first time I saw an alien?

Or maybe, in some sense, they were unrepresentable. Maybe it is only now that these uncanny guests are literally able to appear, to become stickers, film images, patches, T-shirts. Perhaps, even, there is something about the present that forces them to appear, to flush them out of the sky and onto the icon. Do we have any witnesses to this event, this becoming icon of the alien, the alien's abduction by the image? What would prevent a witnessing of such a becoming?

We are called on to produce more than one witness, and in this segment we will be listening to multiple testimonies to this entrance of the alien into our networks of representation. For before it becomes iterable enough to appear, literally, anywhere, the citational image of the alien was networked with a series of singular bodies, bodies that were probed, abducted, instructed, forced to speak. Some of these bodies lived in and off the discourse we call, out of habit, science fiction. Others found themselves on a couch, hypnotized, or speaking out of the pages of a popular book. But all, as witnesses, testify to the sampled quality of abduction.[2]

By sampling, of course, I am referencing the heterogeneous chains of sound and/or image that have composed hip-hop and new music performance for nearly a quarter century. Sampling—the grafting of one sequence of sound or image onto another—relies in its conceptualization and its practice on an understanding of sound and image as information, information whose major effect is found its ability to be networked, connected up to new contexts and moments. By this I do not mean that music, video, or text that samples seeks to "inform" in the CNN sense of the term, although that may also be true. Rather, the emergence of the practice of sampling is associated with the capacity to manipulate sound and images as sequences, a citationality amplified by their digital medium. So, for example, when DJ Spooky operates through the syntheses of disjunctive sound—a track from a stereo album used to test high-fidelity equipment that enunciates the words "stereophonic sound" is blended and refrained through percussion, horns, jet engines, and various styles of noise—he treats the entire audio universe as an immense informatic palette.

> "The Subliminal Kid" moved in and took over bars, cafes and juke boxes of the world cities and installed radio transmitters and microphones in each bar so that the music and talk of any bar could be heard in all his bars and he had tape recorders in each bar that played and recorded at arbitrary intervals and his agents moved back and forth with portable tape recorders and brought back street sound and talk and music and poured it into his recorder array...[3]

As the liner copy suggests, the emergence of sampling as a technique is conjoined with the continual possibility of being sampled. Not merely a manipulative technique of an informatic medium, it describes an ontological condition—the continual and structural possibility of sampling, of one chunk of information being copied and net-worked with another site, "street sound and talk and music...poured...into his recorder array." Indeed, this ontological exposure is registered primarily in the *capac-ities for transformation* complicit with being-sampled. New surface areas of embodi-ment and deterritorialization are constantly exfoliating as technologies of informatic sampling blur the very landscape of "human" consciousness, rendering practices of autonomy, privacy, and propriety into entropic, conceptual formations good only at propagating themselves. Contemporary privacy technologies such as PGP cryptog-raphy and voice encryption seek to reduce the possibility of such media citations—such as a tape recording of a telephone conversation concerning, say, Newt Gingrich and its subsequent high-velocity propagation through the media. Here information economies enable little more than the acceleration and amplification of that most ancient of media: the rumor.

And yet the thought of the informational universe is hardly con-fined in its effects to our usual understandings of "media" or, certainly, representa-tion. They are also an aspect of a larger panic—an encounter with the exobiological character of an informational universe, what funkster George Clinton (the most sampled man on the planet) has dubbed in another context the Mothership Connec-tion.[4] In such an informatic universe, evolution functions less according to a logic of incorporation than an operation of proliferation, a universal ecstasy that thrives on the copy even more than it eats away at an interiority or organism.

For musical sampling is merely one example of the deterritorializ-ing effects of informatics. Moving sounds from one site to another, the high-resolution tweaking of an image sample made possible by morphing technologies differentially repeats transformations in practices of the life sciences. With the emergence of a digital molecular biology organized around the manipulation of traits or alleles as information, the "organism" or even "life" no longer serves as a delimitable object of biological inquiry. Instead—as with DJ Spooky's audio universe—molecular biology operates on an immense recombinant informatic interface. The new citationality of molecular biology enables novel styles of connection and exposure to other traits, other species and other machines. The contours of the human body are being virtually reorganized as the alleged essence of the body, DNA, becomes mobile, a moveable and thus re-moveable script of oneself that is constantly available to an alleged "out-side" that we cannot master.

Pitches are given to venture capitalists, IPOs are floated, living systems begin to coevolve with the stock market. As a result, unprecedented life-forms arrive. It is of course crucial that such life-forms arrive first as nearly transparent commodities available for universal exchange, dissipative structures literally becoming machines for the production of selective wealth. Beyond "speaking" as commodities or transforming our ecology into an allegedly dead zone of things, such novel life-forms occasion new forms of deterritorialization that are quite literally ecstatic, an allopoiesis that overtakes the self-propagation of autopoietic systems. We are, as evolutionary psychologist Merlin Donald might put it, out of our heads, increasingly distributed beings whose cognitive and evolutionary insides are becoming outsourced across a network. Donald wrote of the effect that the allegedly external storage device writing had on the emergence of thought and interiority but he only hints at the forms of consciousness cultivated by a much more amplified and differential infoscape composed by this most recent encounter between capital and evolution.[5]

From Citings to Sightings: Close Encounters of the Informatic Kind

Abduction discourse exemplifies the character of such an informatic landscape in which new practices of citationality—such as the sampling of DNA—emerge. From its very origin, the "problem" of extraterrestrials has been conceptualized as a problem of information. An early (1953) report from the Robertson panel, charged with an evaluation of the UFO problem, concluded that the threat of the Unidentified Flying Object is of an informatic kind:

> Parlous times... result in a threat to the orderly function of the protective organs of the Body Politic. We cite as examples the clogging of channels of communication by irrelevant reports, the danger of being led by continued false alarms to ignore real indications of hostile action, and the cultivation of a morbid national psychology in which skillful hostile propaganda could induce hysterical behavior and harmful distrust of duly constituted authority.[6]

That is, it was the replication and propagation of the rumors and reports of UFO sightings and contact that disturbed the panel, and not the objects of the reports and rumors themselves. Excessive discussion of UFOs could, in and of itself, constitute a very real danger, endangering the organs of the body politic, clogging them like a rhetorical virus. The very contact of Americans with these stories risked the cultivation of a psyche that would be even more exposed and susceptible to a hysterical rejection of authority, a psyche perhaps less conducive to a National Security virus that was so effectively propagating. In short, UFO reports, from the perspective of the U.S. government, were a rhetorical contagion threatening the effective production of narratives about itself, a "duly constituted authority." Such distrust—the belief that their own government's authority was to be distrusted, with the secret truth veiled from the public—apparently provokes *belief in*, rather than *skepticism of* propaganda. The Robertson panel clearly found propaganda to be more contagious, and more "out there," than truth.

Thus the early management of the UFO phenomenon in the United States focused at least as much on the establishment of reporting protocols as on the investigation of sighting events. Allen Hynek, a respected astronomer who worked with several of the early investigations into UFO phenomena, recognized that the ecology of the UFO included its status as a rhetorical practice: "UFOs exist, for most of us, as reports."[7] From very early on, the task of the meager U.S. UFO investigation efforts was to separate the UFO "signal" from its "noise," a task which Hynek would later charge was essentially the denial of the possibility of signal.[8] Indeed, the very code names of early investigative committees—Project Sign mysteriously became Project Grudge—suggested a semio-textual understanding of the alien. Faced with the possibility of a body politic clogged with a surplus of information, investigative teams sought to put the country on a rhetorical diet.

Yet it was the very informatic economy associated with UFOs— a blip on a radar screen, a government report, a "message for mankind"—that would thwart such an effort. For any desire to stifle the news of the UFO will itself make news—witness the recent proliferation of Roswell theories in the face of the U.S. Government's attempt to "come clean." Traditional understandings of the UFO and abduction phenomena have understood the alien presence as a crowd phenomenon, one in which the mass of reports essentially acts as an aggregate, a totality in which one report spawns another which spawns another, proliferating into a mob not unlike the ones that overtook certain sections of New Jersey after Orson Welles's *War of the Worlds* hoax. In this understanding of the appearance of the alien, the observer or reporter is primarily interested in determining what the alien *is*; hence the reports which threaten to clog the organs of the body politic, the univocal messages that they and the aliens bear. In this context, to manage the proliferation of UFO information, one must simply cut it off; in the place of a mass or crowd of sightings and reports, we will have an aggregate of silence. Such a diagnosis, troubled as it is by a "clogging," is consistent with a rhetorical economy of incorporation that characterized the surplus of alien reports to be fundamentally objects of containment.

But perhaps such an understanding forgets the very essence of the alien presence, its characteristic ability to proliferate and mutate, disturbing the various taxonomical categories that we bring to bear on "them." As one psychologist and UFO investigator with a particularly teleological bent put it, perhaps we "are meant to be baffled." While I would challenge the teleological and anthropomorphic attachment of meaning to the UFO phenomonon, it is difficult to argue with the claim that "aliens" and "UFOs" appear to be inherently cryptic; no hermeneutic of disclosure has proved capable of providing sufficient revelation for the growing population of observers, abductees, and debunkers. The discourse mushrooms—with bestsellers, magazines, Usenet news groups, television dramas, and blockbuster films—even as the indeterminacy of the alien phenomenon persists. Many groups claim to know what these phenomena are, but none are sufficiently persuasive to have the final word, so there will still be yet another rerun of the *X-Files*, yet another confession of abduction, yet another debunking.

I will suggest here—and it is my hope that you are suggestible—that in this cryptic context, alien phenomena are better understood as proliferating packs than incorporating crowds, drawing these concepts from Deleuze and Guattari's sampling of Elias Canetti's enormous and marvelous book *Crowds and Power*.[9] For Deleuze and Guattari's Canneti, the (closed) crowd is a mass or aggregate that achieves one voice; its borders are secure enough that one most definitely knows who is "in" the crowd and who is "out" of it. Paranoia—that usual suspect trotted out to explain cold war UFO flaps—is the all-incorporating subject formation associated with the crowd—it is sure that all of these "events" add up to something, even if this something is, in the eyes of the national security agencies charged with the management of the UFO problem, nothing itself. By contrast, pack phenomena are characterized by their continual variation—as with a pair of dice, variation is built into the operation. In a pack, this continual variation—its proliferation, its transience, its subsequently shifting morphology—renders the borders of the phenomenon persistently uncertain. Rather than incorporating everything on its itinerary with the repetitious mantra of "inside/outside, inside/outside," alien phenomena replicate in exuberance, nomadically distributing themselves across diverse and even divergent ecologies.

It is this continual "surprise" value associated with the pack that seems to best map alien events. For as entities that are understood as information—Hynek even calculated what he called the "strangeness rating" of each report, a rating roughly commensurate with Claude Shannon's quantitative description of information—alien phenomena continually surprise observers.[10] One abductee, sampling perhaps from the conventions of *Penthouse Forum*, put it this way: "I never thought it would happen to me..." In this framework, alien events would be understood as cryptic—indeed, as "alien"—not because we "lack" some bit of information but because, as pack phenomena, alien events are characterized by their status as multiplicities more closely associated with swarm behavior than with the understanding of any individual subject. As with the slightest twitch of a starling's wing, the multi-plicities called "aliens" propagate the smallest difference into extraordinary events. Aliens, like a cloud of starlings, arrive.

Alien arrivals, so I would argue, must be understood within the economy of the sample because they are (1) constituted by a citationality that allows them to be iterated within the context of a global hyperspace (I've seen this on TV!) and a corporeality composed of information (They're taking samples of my DNA and selling them to biotech companies!) and (2) because their consistency across all previously instantiated reporting protocols resides only in their variation, like a refrain composed of heterogeneous chunks of sound, an orderly noise continually altered by each successive sample. The crafting of samples demands, above all else, variation, a discontinuity physicist and digital cosmologist Edward Fredkin describes as the "grainy" or discrete quality of an informatic universe such as our own. Fredkin para-doxically describes such a flickering universe where every (literal) *bit* of space/time is binary, one or zero, on or off, as *finite*. Crucial to such a universe is that every bit of space/time is "about" to be something else: the future. "Every part of space is com-puting its future as fast possible, while information pours in from every direction. The result is the same as caused by the apparent randomness of quantum mechanical processes."[11]

This nearly structural fluctuation and uncertainty, of course, gen-erates more and more information associated with alien phenomena, information that seeks to put one final piece in the puzzle and achieve revelation. Many contem-porary abduction accounts—the localization of the blurry tangle of an Unidentified Flying Object and a human subject—narrate and invent the body that would bear such a pack-like excess. Former Van Halen lead singer Sammy Hagar, contacted in 1968 within a dream by an alien presence, told the story this way:

> I woke up and caught the whole thing going down. Either there was informa-tion being programmed into me, or information was being taken out of me to see where my head was at.[12]

This uncertainty regarding the very borders of one's body—not to mention of a dream—is symptomatic of the informatic ecology of the sample. An ecology of con-tinual exposure, the deterritorializing effects of such information transfer erode the very agency of the subject—as sampler or samplee, Hagar's confession emerges out of the space of their convergence, the one who can only say either/or.

But perhaps contemporary subjectivity—*I'm abducted, therefore I am*—is cultivated less out of any certainty concerning one's own agency or the boundaries of one's body, and more out of the continual becoming that the exteriority of information makes available. Perhaps, in the context of an informatic ecology, we need not mourn the panic that emerges from a crashed disk. Such crashes effectively introduce a fluctuation, or what Deleuze and Guattari have characterized as rhythm, into the linear space of binary, serial computation, providing countermeasures to what William S. Burroughs has diagnosed as that other alien virus: language, or, at least, language captured or even abducted by the idiom of the either/or, the Word.[13]

In what follows I will pursue several moments or samples of abduction, samples which will attempt to graft abduction onto an encounter with the capacities of informatic bodies, the bodies we find ourselves living with. Perhaps these bodies—whose boundaries, I have tried to argue, are multiple, massively parallel, at once inside and outside—emerge in association with alien events for good reason: aliens are themselves multiplicities, they come in packs and not individuals. To sample the excessively prolific science fiction author Isaac Asimov, "There is no such thing as a single UFO sighting."

Confession of an Artificial Hoax;

or, The Sokal Construction of Abduction

What do you expect? I am working on one of my innumerable schemes, trying to write up a screenplay that would let me keep one tentacle out of the academy, maybe make some money. The treatment is as follows: It seems that a famous abduction author has disappeared. Now this guy is notoriously fond of publicity—he's been known to actually chase the paparazzi, claim that some celeb has been abducted—so nobody knows quite what to do. The police can't quite take it seriously, I mean the guy's likely to show up any day now, appear out of nowhere in Oprah Winfrey's living room or something. But the days go by, his family is starting to pull their political strings, and a good cop is caught in the middle. He gets pulled off of his current assignment, in which he helps locate missing children, and starts searching for "The Abducted" (my working title) by going undercover into the abduction community, pretending he is an English professor interested in UFO discourse...

Anyhow, that's as far as I get. It's competing with a performance art piece in which I would carry out drive-by shootings of Tamagotchi, you know, those artificial life pets. I want to make a court rule on digital pet slaughter. I could write a book about it, get on Geraldo, maybe even *Lingua Franca*. Oh yeah—they are defunct now. Anyhow, I get sleepy—you know the feeling—so I think I'll close my eyes, just for second, you know, rest my eyes from the screen, nothing major.

Next thing I know, its one of those hypnagogic surf expeditions. I'm asleep, I *know* I'm asleep, and yet my body feels like it's on a mission of its own, trying to tell me to wake up. My body's a field of electrical forces, a body online, and unfortunately my vocal cords seem to have hung up on me. I usually try to yell myself into wakefulness in these situations, but no dice. My mouth opens and closes, but nothing but a wheezy breath results.

All of a sudden—and this is the weird part—I'm thinking about, of all things, Alan Sokal. Now I couldn't care less about the whole science "wars" thing—I mean, I know the academy needs its fair share of the scandal space in the mediascape, but these days thought alone would seem to be enough of a scandal. Do we really need border wars? How mid-twentieth century, this concept of the border...

Anyhow, I am overwhelmed by this incessant, monotonous thought: "Alan Sokal. Alan Sokal. Alan Sokal." So I'm thinking, yes, I should have thanked him. I should have thanked him for citing me in his *Social Text* article. Those of us in the pre-tenure orbit need all the citations we can get, so I should have thanked him.

"Alan Sokal. Alan Sokal. Alan Sokal." Thank You Thank You Thank You Thank You, I'm thinking. Thank You, already. But the rhythm is un-interrupted. I have heard that dreams have porous borders, fractured enough that other people find themselves sharing a dreamscape, like a toaster that picks up a stray radio signal from a passing taxicab or a cop. New York is only about a four-hour drive, which is very little at the speed of dream, which is no doubt electromagnetic.

But it gets worse. My humming electronic body suddenly decides to become as moveable as the information that composes it. In the abduction litera-ture, the verb is "to be floated." It's the common method of transport for bringing those pesky human bodies into the proximity of the alien vehicle. So I'm floating out through the windows that I was sitting beside, going, at the very least, somewhere else. I wave at Amy, my sleeping partner in crime, wishing we had a camcorder.

I should have seen this coming. John Mack's remarkable book, *Ab-duction*, seems to have an unusual number of subjects drawn from rural Pennsylvania.[14]

I don't know what is happening. It seems like all of a sudden I am in a large, rather sterile room. Maybe I am in a faculty meeting! Then I see the beings—I recognize them from television. I immediately think of how insulting it was to call them beings. Hadn't our own philosophical tradition been saddled enough by the monotony of being? There are, I think, four of them, but every time I turn my head I see something else, so who knows? There are, as the mathematicians like to say, *n* of them. I ask "them" what they want to be called.

"Alan Sokal?" someone says with that questioning inflection such as the one I will end this sentence with? "Oh no," I'm saying, "no you've got the wrong guy. I'm rhetoric boy. You want physics boy." I'm gesturing, scratching my head, pointing, giving them directions. "He's in a different neighborhood altogether. Head east on I-80, make a right at the Lincoln Tunnel, head for Washington Square. I don't have his number, but I'm sure if you contacted NYU, you..."

"Alan Sokal?" Okay, so now this one—I'll call them, after *Communion* author Whitley Strieber, visitors[15]—this one visitor is "sounding" kind of desperate, giving me a look that I can only characterize as pleading. Is this some sort of kinked alien fantasy where they want me to play at being Alan Sokal, some kind of role-playing gig where they hijack organisms and force them to act like physicists who act like postmodernists? No wonder they need to wander the galaxy, I'm thinking. These are some rare tastes...

"Alan Sokal!" This one is now gesturing, pointing with one of four digits at a small table. I wonder if having only four digits encourages a species to produce a more effective external memory device than writing. I mean, it would do fine for helping to remember area codes, they're only three numbers, but what about phone numbers? Or doing your multiplication tables over five...'Course there are toes, I wonder how many toes...

Okay, now I've done it. A visitor is waving a book in my face. No doubt my name is inscribed in it, all my minor, pathetic crimes are detailed there. "I'm sorry!" I shriek. Then I notice the book. It's in English. The light isn't so great in here so I'm squinting. It says: *Social Text*.

"Alan Sokal? Explain. Please."

Okay. Now I get it, I think. They're asking *about* Alan Sokal. Like they were in the neighborhood, you know, and they want to discuss these fascinating science wars. What, do I run the local Mediocre Books reading group or something? I want to tell them to give it up, read something good like Kafka or *Melrose Place* and get out of here, but they have come all this way..."Gather 'round..." I say, like some scout leader beginning a ghost story...

"Alan Sokal is a celebrity. He got famous, for a while, claiming to understand poststructuralism even though he didn't. Perhaps more importantly, he claimed to demonstrate that it didn't matter if he understood poststructuralism or not. He conducted an experiment in which he tested the effect of *pretending* to understand some concepts to see if that pretense would be sufficient to materialize a text—a manuscript—into a *Social Text*. It did. He passed the Turing Test of science studies—somebody who was touring through science studies couldn't tell if his manuscript was real or not."

No laughs for the pun. A tough crowd.

"Anyway, from this he concluded that the things he pretended to believe in didn't mean anything. If you ask me—which you have, sort of—what he demonstrated was something much more interesting: the fact that pretending to believe a concept has a material effect on that concept: It helps get it repeated."

"You say it helps to get it repeated. Alan Sokal's belief is very powerful?"

"No, not particularly, although it did help that he was a physicist. No, more important than Sokal's particular belief was the fact that his pretense enabled his concepts to be repeated, iterated."

"Alan Sokal?"

"I said that by pretending to believe in what he wrote, it got written. How else, you might ask, could he have written it? It's clear enough that this was true of the transformation of the text into *Social Text*—they presume, in some baseline way, that author's "believe" what they write."

"But Ross said . . ."

"I know, I know, it could have been a parody, but the principle is the same, even if Sokal had never published it: Only that which can bear repetition is writable. And to be repeated, it of course must have at least the pretense of existence—it must seem to have already been repeated once, if only to oneself."

"Could you say that again?"

"No, I mean why bother? I am just citing an essay from somebody that Sokal likes to make fun of. What Sokal demonstrated, quite brilliantly, was the essentially iterative character of writing. Oh, he may have demonstrated that some members of the science studies pack don't know how to deal with the effects of simulacra very well—which is pretty remarkable, considering how fascinated we are by them—but above all Sokal demonstrated the essential fact about writing: its ability to be sampled, to be cut out of one context and sewn into another. By writing an essay whose primary function was to be misread, Sokal highlighted the continual capacity of texts to mean something other, to become something else. This is not a cheap pluralism—let a hundred interpretations bloom—but an empirical attribute of writing: 'This force of breaking is not an accidental predicate, but the very structure of the written.'"[16]

The visitors look at each other. There is much murmuring, some-thing that sounds like laughter inside my head. It is not unpleasant. I am feeling a little more comfortable—I'm the one with expertise here. It doesn't seem like I am in for any impromptu genetic research, so I draw on my training, turning the tables with a rhetorical flourish...

"Just out of curiosity, why didn't you ask Sokal about all this?"

"We weren't sure if we could believe him." More "laughter." This visitor had a point. Another one comes forward—oh, I recognize this one...

"Could you explain this 'hoax' a bit more? If Sokal thought that these 'concepts' of 'poststructuralism' didn't mean anything, why didn't he write a message that *contained* this information?"

"In some sense, he *couldn't have*. As you know, the information value of any message is essentially its surprise value."

"Claude Shannon, yes?"

I somehow am not surprised that the visitors had gotten the mes-sage on Claude Shannon. "Who would have been surprised if a physicist with a Marxist past decided to critique science studies? Perhaps more crucially, a critique as thorough and as effective as Sokal's would have taken up hundreds of pages over a period of years. And he probably would have had to work through one of the most sluggish information transmission mechanisms known to the multiverse—something called a 'university press,' a media outlet in which one is lucky—or unlucky, as the case may be—to be read at all, let alone widely. By engineering an effective hoax, Sokal pursued a very efficient strategy, informatically speaking: he compressed his message to a smaller—if not the smallest—sequence of characters. This compression also increased the velocity of the message. According to at least one of my sources, news of a hoax actually beat the issue of *Social Text* to the stands. A hoax, in this con-text, is therefore a rhetorical strategy that parasitically uses the pretense of 'belief' to reduce the time and information necessary to propagate a message. 'Belief' is like a compression algorithm, then.

"Crucially for Sokal's strategy, though, his hoax was dependent upon others. He could never be sure, in advance, that his message would be trans-mitted, that his algorithm would run effectively. There was no way to cut to the chase, as it were—he just had to send the message and endure the waiting; his message needed to be actualized. Paradoxically, in order for his message to be completed, *the recipient must not understand it*. Indeed, unlike our situation with the university press, *it is best if no one reads it at all*. Hoaxes such as Sokal's are not simply acts of deception, although some hoaxes are—such as counterfeiting. Rather, hoaxes of the order of Sokal's operate through the difference between the initial reading of a message—even if it be a nonreading—and a subsequent one, one that the hoaxer provides. We'll talk about that problem later."

"Later? What is a problem later?"

Yes, I remember from Betty and Barney Hill's 1961 abduction—a breakthrough case—that some of the visitors have trouble with time.[17] Well, we've got that going for us, that ingenuous dicing up of time into homogenous, repetitive chunks, one grafted onto another and another...

"Nothing. It's *now*, I guess. What I mean is that the hoaxer sometimes has a hard time establishing credibility when they finally come out with what they claim is the truth. Philip K. Dick—"

"Yes, yes, we know *him*," they all chanted in unison, with much laughter.

"Philip K. Dick drew this logic out to its remarkable conclusion in a book called *A Scanner Darkly*.[18] A man appears on a television show as a world-famous imposter. He was a real scammer—he'd impersonated a doctor, performed brain surgery, argued cases before the Supreme Court, impersonated a lion tamer—he'd done it all. Of course it turned out that the guy hadn't done any of this. He was just a janitor from Disneyland, pretending to be a world-famous imposter, which, of course, he now was...

"So imagine if Sokal had printed up some counterfeit money—good hundreds—and passed them at a bank, planning to come in later and say: 'Well, I hate to tell you, but I gave you some phony money earlier today. It doesn't seem that your detection system works so well. It doesn't seem to me that this whole monetary system has any merit; it can't tell the difference between a real representation of value and a bogus one.' I'm not sure he would have done so well with the local prosecutor. And the prosecutor would have a point—why should we believe a counterfeiter?

"But Sokal had a very minor version of this problem. Not much money was at stake, I guess. There were apparently some parts of his essay that he was 'serious' about, according to the short piece he wrote after the hoax was revealed. So, for example, Sokal had some interesting things to say about the ways in which mathematics is treated in our school curricula. So there were components of the essay that somehow escaped the status of hoax. Some of it—besides his name of course, although that too came under scrutiny—composed a message that was designed to be understood, even if only on a third or nth reading. That is, one doesn't—from Sokal's perspective—understand Sokal's arguments concerning pedagogy the first time around, because they are nested inside an enormous forest of citations. I can barely keep track of who's talking; it's difficult to figure out when Sokal is speaking and when he is citing somebody. This appears to be deliberate rhetorical tactic—argue from authority, graft the text onto innumerable other ones. On the 'second' reading, of course, one doesn't believe a word of it—it's all syntax in the service of a hoax, with the only semantics residing in its status as a fake. 'What's this garbage about pedagogy?' we say with a knowing smile. 'I'm really surprised Ross wasn't tipped off by that bit. I mean really, he can't be serious . . . '

"Oh, but then we read his commentary on his own text, and we learn that he was serious! So only now do we engage with the argument that Sokal had constructed: math is being taught poorly *and* all this pomo crap is doing nothing to help it. Not bad, actually, if a little off topic."

Sage, concerned nodding from the visitors.

"But you have to think that now there is no a priori reason why there might not be other 'serious' parts. At the same time, he might not be serious at all about this pedagogy stuff. I mean if he's so serious about it, why distract attention from that part of the essay by making it true?"

"I like to pretend that his footnote to me is serious. I mean, after all, it would seem that he was engaging in a canny, pomo bit of doubling in actually citing me, since the first line of my cited essay is concerned with citationality itself: "We have been inserted into a technoscientific text."[19] Yes, I feel a strange complicity with Alan Sokal's hoax. The primary rhetorical operation of his text is citation, the cutting of text from one place and its insertion into another—the footnote. What could it mean that he cites me writing about citationality? If he is not serious in citing me, he sure chooses a funny way to show it: citing me is to refer to the importance of the operation of citing, and in citing, executes the content of the essay. We're back to the world-famous imposter: Sokal pretends to cite me about citationality. It turns out he was doing no such thing—just piling up my citation with the other pomo nonsense. But in piling up my citation about citationalty, he cites me. Indeed, my essay—on the third, posthoax reading—actually cites its own citation into the essay. The fact that the essay is only referenced within a footnote and not in the body of the text only amplifies the reference to citationality."

"Citationality? This is our method of propulsion, no?"

The one with the copy of *Social Text* is speaking up again, shaking the journal in my face. I knew it. The visitors had, unbelievably, learned something from Sokal's alleged hoax on quantum gravity. Against his wishes, Sokal had made some sense. Maybe about wormholes.

"Oh." I am on now, really on. "You mean that the traversal of a wormhole relies, as Derrida would put it, on the essential disseminatory character of the universe, its time out of joint? Yes, that's it! For Derrida—the Heraclitus of discourse—one never speaks out of the same mouth twice. Indeed, the very concept of the 'same' mouth forgets its multiplicity, its eternal variation, its *différance*, its difference from itself. Now I know why I went to graduate school! With wormholes, information and material gets inserted into one 'mouth' of the wormhole and emerges from another 'mouth' at another time." Here, I cite wormhole expert Kip Thorne:

> A wormhole is a hypothetical shortcut for travel between distant points in the universe. The wormhole has two entrances called "mouths," one (for example) near earth, and the other (for example) in orbit around Vega, 26 light-years away. The mouths are connected to each other by a tunnel through hyperspace (the wormhole) that might be only a kilometer long. If we enter the near-Earth mouth, we find ourselves in the tunnel. By traveling just one kilometer down the tunnel as we reach the other mouth and emerge near Vega, 26 light-years away as measured in the external universe.[20]

"From Derrida's perspective—which as far as I know would be included in the external universe—such an example of time travel is not one example among others: it allegorizes the essential dislocation at play in the emergence and operation of writing. The attempt to articulate the becoming space of time, the becoming time of space necessarily—and this would simply be the necessity of chance—has recourse to the very medium of its enunciation—the primordial dis-location of the mouth, the voice's wandering away from itself, eternally varying such that one cannot rigorously determine the boundary of one mouth and another. Indeed, Saussure's famous diagram of the sign—two mouths connected by dotted lines to signify communication—could provide us with another 'mouth' here. This is all the more perfect since Kip Thorne's work on wormholes itself emerged out of some consulting work he did in the writing of a novel by Carl Sagan—that is, Carl Sagan called up and asked Thorne if he could work up a scheme that he could cite, and after a few thought experiments Thorne provided Sagan with a method of transport for his novel, a scientific description of time travel through worm holes grafted onto science fiction.[21] This is what you meant, right?"

"No, negative, not our thought," the visitor says, "although all of that is very interesting." They are inscribing this stuff on glowing, glyphic tablets. The glyph looks like something like a fish.

"What I mean is that we, like your physicists, gleaned the concept of the wormhole from science fiction, an *Amazing Story* from the 1920s that we found in an archeological dig on our planet. *This is* science fiction, no?"

Close Encounters of the *n*th Kind

The universe, I realized, was being turned inside out—reversed. It was an eerie feeling, and I felt terrible fear. Something was happening to me, and there was no one to tell it to.

Philip K. Dick, *Radio Free Albemuth*

True, Philip K. Dick took a lot of drugs. That much we know. Author of more than thirty-six novels—mostly science fiction but some that are sole members of a class consisting of themselves, such as *Valis*—Dick was a skilled high-speed typist and a brilliant, sardonic, and hilarious weaver of narrative. According to published accounts of his own journals—an eight-thousand-page, unnumbered journal entitled "Exegesis"—he fell into somebody else's novel: "I think I read all this in the novel THE ROBE x years ago. Jeez, I've fallen into someone else's novel!"[22]

Such a falling, of course, sounds familiar enough to viewers of clay animation hero Gumby, who could "jump into any book with his pony pal Pokey too," but Dick's experience was something other than a leap into the adventurous world of textuality. Indeed, for Dick, the exposure to what he would later dub "Valis"—the Vast, Active Living Intelligent System that spoke the "hideous words" and sentences oozing from the radio—forced nothing less than an encounter with the informatic character of the universe.

This encounter was itself multiple. Dick wrote at least four novels that are implicated in what he would call "the events of 2-3-74," a mysterious laser-like pink light, a "firing of information" at his head that took place in February and March of 1974, after his exposure to a glyph of a fish worn around the neck of a pharmacist delivery woman at his apartment door. Each treatment of the event swerved in its own peculiar fashion from a "representation" or "decoding" of the information that may or may not have been fired at Dick's brain. The novels and the Exegesis—itself formed out of pages that constitute non sequiturs of each other, as there is no rigorous sequence to be had—form less an unraveling of Dick's mysterious encounter than its repetition, a proliferation without origin. *Valis*, for example, copies elements of the Exegesis's treatment of 3 February 1974 into its narrative, a narrative in which Horselover Fat's story is told by a science fiction writer named Phil. "Fat," of course, is "Dick" in German, and *Valis's* narrative operates less as the "expression" of Dick's story than of its grafting into a fictional universe that, in some sense, differs from itself. So, for example, Phil tells us that Valis informed Fat of his son Christopher's health:

> He knew, specifically, that his five year old son had an undiagnosed birth defect and he knew what the birth defect consisted of, down to the anatomical details. Down, in fact, to the medical specifics to relate to the doctor...His brain had trapped all the information the beam of light had nailed him with, but how could he account for it?[23]

This would of course suggest to the reader that this text describes the event endured by Dick. Dick, too, we learn in the Exegesis, interviews, and other novels of the *Valis* trilogy, knew the precise characteristics of a previously undiagnosed birth defect after the burst of information. It probably, Dick claimed, saved his son's life. This reading is encouraged by Dick's citing of his own Exegesis—what Phil the narrator characterizes as a "journal" but what Fat calls his "exegesis"—into the text of *Valis,* a text that testifies to the informatic character of the universe:

> One of the paragraphs in Fat's journal impressed me enough to copy it out and include it here. It does not deal with right inguinal hernias but is more general in nature, expressing Fat's growing opinion that *the nature of the universe is information.*[24]

In contrast to the precise information "transferred" into Fat/Dick's brain, however, the copy or citation of the Exegesis operates through its difference from itself. *For the Exegesis cited in the text of Valis is not the Exegesis at all*—according Lawrence Sutin, editor of those Exegesis selections published as *In Pursuit of Valis,* it appears nowhere in the eight thousand pages written by Dick in his "apology for life."

> While the tractatus does recapitulate a number of prominent theories posed in the real-life Exegesis, it is a separate work—polished distillations intended to fit within the framework of the novel—and not a genuine selection of quotations from the Exegesis itself.[25]

What is at stake here in the Exegesis's citational difference from itself? While Sutin immediately takes recourse to the notion of the "genuine"—itself a category continually and hilariously evacuated in Dick's novels such as *Man in the High Castle* and

A Scanner Darkly[26]—it would seem that what is cited is not the Exegesis at all, but the difference between the Exegesis and its citational graft, in other words the difference between the Exegesis and itself. What is cited, then, is the continually differing character of exegesis, a differentiation that would thwart any final understanding of the event, except an understanding of the event as itself a form of constant variation—Edward Fredkin's "grainy," discontinuous universe, where a change in kind is always about to happen: the future. Fredkin's second law of information reads "an informational process transforms the digital representation of the state of a system into its future state." The fictional citation of the "Exegesis" articulates this as well as can be expected:

> Thoughts of the Brain are experienced by us as arrangements and re-arrangements—change—in a physical universe, but in fact it is really information and information processing that we substantialize.[27]

Change, differentiation, is articulated here as operations of information, operations such as citing one text into another, a process of sampling which itself allows for the articulation of an event's difference from itself. Just as Dick deploys the alter ego "Horselover Fat" marking the difference between author and autobiographical character—that is, a difference between Dick and himself—the sampling of a simulacrum of the Exegesis dramatizes the variation that is the core of the exegetical practice. If exegesis is that textual, interpretative practice devoted to disclosing the truth of scripture, then Dick's operation of exegetical citation discloses the truth of February/March 1974's continual variation. The only way for Dick/Fat to testify to the pink light was to constantly change his story.

The Exegesis, then, is "grainy," continually differing from itself as it is made of variations and discontinuities. Such proliferation, though, paradoxically encourages a cosmology of epistemological finitude. True, such a digital, informatic universe would seem to be a deterministic one—even Fredkin argues that such a universe is finite to the extent that it is a "closed" system, leading him to postulate the possibility of simulating moments "before" the big bang, a logic of reversibility that disables any understanding of the universe as an incorporating "inflationary" or "Pac-Man" cosmos whose inside continually engorges itself until collapsing in on itself. But "the deterministic nature of finite digital processes is different in that it is unknowable determinism." Fredkin ties this to Turing's *entscheidung* problem, which hamstrings us concerning the binary prediction of any particular program's instantiation. To halt or not to halt? There is a "celebrated proof that, in general, there are no analytical shortcuts that can tell the future state of some general computation any quicker than doing the computation step by step." That is, computation is epistemologically finite; its ability to predict is limited by the speed of available computation. As Fredkin puts it in a footnote, "In general, physics is computing the future as fast as it can."[28]

Hence rather than attempting to understand the meaning of his informatic abduction, Dick's rhetorical practice of exegesis was an ordeal of endurance, a capacity to instantiate information. There is comedy and exhaustion in this cosmic, albeit machinic, consciousness that requires continual interaction. What was at stake for an identity in contact with Valis is survival and not knowledge, a survival that paradoxically induces a transformation.

Valis is the medium is the message. More crucial than the content of the information fired is the suddenly affective encounter with the possibility of living in a universe of immense and entangled complicity, where Fredkin's universe of information operates all the way down to twisted strands of DNA. Such a universe is one of total exposure—the opening of a door leads to a burst of information, information that itself emerges only from "elsewhere," without origin. In Dick's formulation in the Exegesis, such information is not omnipresent but "multipresent," always, possibly, about to erupt, to appear around the next corner. "Could it be that in 3-74 Christ interfered with the genetic coding that had programmed me to die at that time? . . . My gene-pool (DNA) memory fired—opened up: I know that."[29]

This continual exposure is not, as in the familiar paranoid formulation, peculiar to Dick in his cosmology; the encounter with Valis provoked Dick to claim that the universe is itself a continually variable informatic field: "We appear to be memory coils (DNA carriers capable of experience) in a computerlike thinking system."[30] Dick, his characters and his readers, to the extent that these can be rigorously distinguished from each other, dwell in an informational universe for whom there is no cutting to the chase, no ethical or ontological principle capable of knowledgeably preparing them for a fundamentally differential identity, future.

While this emphasis on the machinic, informatic character of Dick's universe may appear to be a symptom of a writing practice that included *Do Androids Dream of Electric Sheep?* and a drug routine consisting of handfuls of amphetamines washed down with milkshakes, we have recourse to another archive that suggests Dick's understanding of the continual exposure fostered by the increasing prevalence of ecologies of information was enmeshed with an equally material set of events. In addition to being fascinated with the role of taping technologies in the ouster of Richard Nixon—an event Dick claimed retroactively to have foretold in his 1970 manuscript *Flow My Tears, The Policeman Said*—Dick was the object of a mysterious informatic abduction in November of 1971. His Marin County home was broken into, and his fireproof archive of files and work was ransacked and exploded. Local law enforcement was less than energetic in their investigation of the crime, and over the years Dick continually speculated about the identities of the perpetrators that stole and sampled his archive. Indeed, at one point his list of suspects actually included himself. In a 1972 letter found amongst his Exegesis papers, Dick writes:

> I blew up my own house and forgot I did it. But why did I forget that I did it? So I'd think that I had an actual enemy so I wouldn't have to face the fact that I'm paranoid, i.e. crazy. I blew up the house to convince myself that I was sane. Anyone who would go to that much trouble must really be nuts . . . [31]

The "truth" of Dick's informatic abduction, though, is beside the point. Not because, as some in the abduction community would have it, it is the experience of the abductee or experiencer that is central to the abduction event. Instead, what Dick's experiences with Valis narrate brilliantly is that the encounter with an "alien" intelligence is not about a message that is being given to earth or the experiencer. For the possibility of parsing or consuming such an experience into a message depends upon the capacity to rigorously compress the interpretation of the event. As Dick's relentless encounter with the shifting boundaries of his own experience indicates, such a compression is not possible precisely because the experience of the alien is an encounter with the exteriority of information—the fact that information, to paraphrase Socrates's rap on writing, is always possibly getting into the wrong hands, always possibly going awry or being forgotten.[32] Indeed, to tell the "truth" about exteriority is precisely not to rigorously distinguish the true from the false, because real information wanders into fiction and vice versa—hence Dick's "forgetting" of his own detonation, his "fall" into someone else's novel. Or maybe his own: "I seem to be living in my own novels more and more . . . My books are forgeries. Nobody wrote them."[33]

Dick's encounter with Valis, then, was less about the interpretation or ingestion of a message than the execution of an algorithm, an algorithm of varying commands to differ, to sample or be sampled. In Dick's phrasing, only an entity that "would be itself and not itself continually" would be adequate to the experience of Valis: "I almost became a sincere tool of a conspiracy consisting of myself." Such continual variation would for Dick, in principle, extend not only to the self that "bore" the event but also to the event itself, whose retrospective veracity could never be separated from its profound difference from the present. Valis uttered impossible sentences whose very effects provoked the inability to differentiate between the real and the unreal, thus recoiling on the status of Valis's statements themselves.

> In one of my novels . . . certain anomalies occur which prove to the characters that their environment is not real. Those same anomalies are now happening to me. By my own logic in the novel I must conclude that my (own) perhaps even collective environment is only a pseudo-environment . . . as in the novel I can't figure it out. It makes no sense.[34]

Any narrative of Valis—a testimony to what Valis was—paradoxically entails a memory not of the past but of the present: "I claim to remember a different, very different, present life." This other present speaks to a second register on which the experience of abduction fosters something other than a message or understanding of a truth—rather than a meaning that is decoded or interpreted, the experience of the alien provokes transformative effects, becomings that Fredkin might dub the future:

> I have been transformed, but not in any way I ever heard of. I drive differently, much faster, reaching for an airvent on the dashboard that is not there. Evidently I am used to another car entirely. And when I gave my phone number the last two times I gave it wrong—another number. And to me the weirdest thing of all: at night phone numbers swim up into my mind that I never heard of before. I'm afraid to call them; I don't know why. Perhaps in some other part of Orange County someone else is giving my phone number as his, drinking wine for the first time in his life and listening to Rock; I don't know. I can't figure it out. If so, I have his money . . .[35]

In some sense, then, differentiation extends to more than the interpretation of the Valis event—it extends to the transformation of Dick himself. For Dick, the universe has ceased to be an outside that he somehow lives "in"—instead, the "memory coils" called Philip K. Dick are in continual contact with the very surface of the universe, information that continually transforms him. "I feel I have been a lot of different people." Philip K. Dick composes a *Möbius* body, a corporeality both "in" his novels and just plain out of it, an interior subject called a "writer" who is nevertheless continually exposed to the strange hacks of Valis, who may or may not exist. Indeed, as in the example of Sammy Hagar with which we began, Dick's subjectivity and his practice as a writer emerges precisely out of the impossibility of distinguishing an inside from an outside.

Gilles Deleuze, in his 1968 book *Coldness and Cruelty*, argues that Sade and Sacher-Masoch are among the great clinicians of medicine—they each isolated the affects and orientations that cultivate their eponymous syndromes. Sacher-Masoch, for example, excelled in the differentiation of masochism from an alleged "love of pain," focusing instead on the masochist's penchant for waiting and anticipation. In short, Sacher-Masoch shows the masochist to be cultivating a capacity for the future, a future of "a new man" who has removed the image of the father and made possible a Nietzschean overcoming of the human.

Although Deleuze's claim concerning Sacher-Masoch's singularity is open to question, I would nonetheless like to apply his understanding of the clinical contribution of literature to Dick's treatment of abduction. For while the psychological, historical, and bestselling authors of abduction discourse tend to emphasize or at least engage the "truth" of abduction scenarios, Dick isolates what is perhaps most singular in abduction narratives: their capacity to proliferate and induce transformation in diverse ecologies, to instill what Deleuze and Guattari have characterized as "becomings." Rather than mimetic stagings or performances whose goal is representation or completion, becomings are continuously variable events organized around aspects or "particles" of an object or an event. "What is real is the becoming itself, the block of becoming, not the supposedly fixed terms through which that becomes passes."[36]

Nor does becoming-wolf, for example, entail the ingestion by or of the wolf: "All the better to eat you with," says the alien. It is this characteristic of being on the border, both inside and outside, a Möbius space, a space of impossible consumption, that characterizes becomings. These continual transformations operate not only at the level of a subject—"I feel myself becoming a wolf"—but of multiple subjects—a couple of authors, a pack, a polis. In constant motion, the pack, the alien, is always somewhere in particular, but it has n centers. You can locate the center or "leader" of a distributed pack when you find that which never ceases to differentiate—that which differs from itself, which will, of course, always be elsewhere, later.

It is this encounter with the informatic and therefore variable uni-verse that marks Dick's relation to "Valis," an engagement whose defining characteristic is its continual difference: Valis always has the status of n, unknown. Indeed, knowledge, in the usual sense—what was it?—is hardly material in Dick's relation to Valis. Rather than knowing what Valis "means" or what it "was," Dick engaged the incessantly experimental and profoundly ethical question: *What should I do?*

Astronomer Allen Hynek coined the by now familiar taxonomy of the alien: close encounters of the first, second, and third kinds.[37] Since his death, a fourth category has been added: the encounter of the fourth kind, abduction. Dick's experiences merit yet another strata of the encounter with the alien, one that perhaps disrupts taxonomy altogether, as it engages with an encounter of the *n* th kind: the encounter with alien thought, the thought of an alien as itself a *consequence* of the informatic character of the universe. Perhaps, even, Dick's encounter with Valis teaches us that, in the alien encounter, one precisely comes into contact with thought, not a narcissistic discovery of ourselves in the cosmos, or even an understanding of the finitude of our knowledges, but the incessant, asignifying exteriority of thinking, the sampling and being-sampled that thought entails. To think, sampling Deleuze narrating Foucault, is to fold, to double the outside with a coextensive inside. Thinking is always in contact with an outside, a distant other who is bent into subjection, but not without haunting interiority with its proximity to a monstrous alterity. To arrest the arrival of the alien with the question of ontology—"Ok, What *is* it?"—or epistemology—"Prove it!"—is to enter into the game of lack, an infinite game in which neither the negative case—"The alien does not exist"—nor the positive—"Finally, conclusive proof that Christopher Walken is from Proxima System! Film at eleven"—takes us anywhere.

But the emphasis on the epistemological and ontological character of the abduction event does more than mire conversation in a stagnant differend, an agon that fails at resolution in an analysis of very real becomings. It also overlooks the very possibility of any event whatsoever, the cultivation of exteriority that is, among other practices, reading. Psychologist John Mack characterizes the effects of abductee testimony in terms of development and evolution, but stripped of its teleology. We might also link such speech acts to "involution" and "deterritorialization," operations of folding and the emergence of heretofore "alien" connectivities:

> Stated differently, the information gained in the sessions is not simply a remembered "item," lifted out of the experiencer's consciousness like a stone from a kidney. It may represent a developed or evolved perception, enriched by the connection that the experiencer and the investigator have made.[38]

For it is in a matrix of hospitality, an exposure to an other, Fredkin's unavoidable, finite, "running" of abduction events—the encounter of an *n*th kind, the informatic permeability of thought—that abduction itself gets abducted, taken for a ride, carried away. Getting into the wrong hands, abduction becomes something other than its genealogy would suggest, a technology of identity that emerges out of an encounter with the unthinkable, a shattering of thought that disables the reproduction of any interiorized identity. Taken by surprise, abduction enters the house of being and squats, deterritorializing the household with its *unheimlich* effects.

Ironically, abduction discourse tends to overlook the very vehicles of abduction, the materiality and subsequent rhetoricity of the Web sites, books, films, government documents, hypnosis sessions, and case studies concerning abduction. On one hand, this overlooking of the work that the reading and viewing of abduction carries out for the alleged consumers of abduction narratives is far from containing any surprise for us: the very overlooking of corporeality fostered by the rhetorical ecology of the informatic body forms a discursive fractal with the elision of abduction discourse's strange rhetoricity: its mode of delivery, its complex of affects, its delays and amnesias. The abduction event is itself, like its discursive cousin DNA, conflated with its writing, swallowed by the Word.

On another tentacle, however, this forgetting bundled with the informatic body, this interrupting of the flesh, marks precisely the dislocation and variation at the heart of these informatic bodies. Informatic bodies are composed of, precisely, surprise. The becoming-information of the body, rather than miming an alleged disembodiment of informatics from the "flesh-eating 90s," provokes a double take, and all the problematics of doubling, on the topologies of human subjectivity. Amnesia, for example, opens human experience to the charge that it can't remember itself, that it is forgetting itself in its becoming-inhuman. This forgetting becomes, then, a responsibility, a demand for testimony to the past. It also becomes a new capacity, a response-ability, the capacity to welcome an other who has surprised us.[39]

Notes

0. Welcome to Wetwares™, N.0

1. Lynn Margulis, Dorion Sagan, and Philip Morrison, *Slanted Truths: Essays on Gaia, Symbiosis, and Evolution* (New York: Copernicus Books, 1997), 17.

2. Ilya Prigogine and Isabelle Stengers write of the "dialogue" that is the scientific encounter with nature. Such a model does much to amplify the fundamental complicity of scientific questioning with the very obser-vation and apprehension of objects that are other to it, as dialogues are composed out of a fundamentally uncertain *response*. "Dialogue" requires more than partners exchanging bundles of information, but instead involves a differential commons that makes available not consensus but an irreducible exposure to the other. My description of the technoscientific scene of artificial life as one of seduction suggests that in some instances this dialogue arouses more than a transfer of signs or *logos* and becomes an asignifying practice of pathos, eros, joy, wonder, and panic. See Ilya Prigogine and Isabelle Stengers, *Order out of Chaos: Man's New Dialogue with Nature* (New York: Bantam Books, 1984).

3. Stefan Helmreich, *Silicon Second Nature: Culturing Artificial Life in a Digital World* (Berkeley and Los Angeles: University of California Press, 1998).

4. Gilles Deleuze and Félix Guattari, *Anti-Oedipus: Capitalism and Schizophrenia*, trans. Robert Hurley, Mark Seem, and Helen R. Lane (Minneapolis: University of Minnesota Press, 1983), 36.

5. Ibid., 37.

6. Gilles Deleuze and Félix Guattari, *A Thousand Plateaus: Capitalism and Schizophrenia* (Minneapolis: University of Minnesota Press, 1987), 395.

7. In *The Western Lands* (New York: Penguin, 1987), William Burroughs writes as if his protagonist Kim Carson were a hacker deploying a particularly itinerant weapon during his time in Mexico City. "Kim finds work in a weapon store and devises variations on the Maquahuitl . . . obsidian chips set in wood, the usual shape being rather like a cricket bat . . . long whips of flexible wood. . . . They cut and break on bone and every break is a new cut, a new edge" (70–71).

8. The scare quotes around "in" mark my ecstatic fear that "past" and "present," "inside" and "outside" are topological folk psychologies incommensurable with the practice of autopoietic emergence. It is the arrival of the Outside—what both Charles Sanders Peirce and much contemporary ufology characterize as abduction—that perhaps best maps such an event, but only in so far as such arrivals may crack discourse itself open into pro-liferations of holes, cracks, and fissures provoked by novel encounters with exteriority. Poet Jack Spicer wrote of such a poetics of disciplined perforation or "dictation" from the Outside while Burroughs sum-moned the capacities of the Third Mind through the caring slashes of the cut-up.

9. Stuart Kauffman, *The Origins of Order: Self-Organization and Selection in Evolution* (New York: Oxford University Press, 1993), 227.

10. William S. Burroughs, *Place of the Dead Roads* (New York: Holt, Rinehart, and Winston, 1984), 67–68.

11. Ibid., 68–69.

12. Ibid., 52.

13. Deleuze and Guattari, *A Thousand Plateaus*, 400.

14. Burroughs, *Place of the Dead Roads*, 27.

15. "He decided to call these beings by the general name of 'familiars,' which is a term usually restricted to animals. They were certainly familiar and, like animals, attempted to establish a relationship with a human host." Burroughs, *Place of the Dead Roads*, 168.

16. Burroughs, *The Western Lands*, 29.

17. Crucial to these qualitative events of seduction—becoming-familiar, becoming-gun—is their debt to an essential multiplicity: the refrain, the organization of repetition by meter or rhythm. Deleuze and Guattari report (*A Thousand Plateaus*, 311) that the refrain has at least three movements or components in its algorithm: (1) *Invent a Surface.* "A child in the dark, gripped with fear, comforts himself by singing under his breath." Here breath acquires a plane, a space with an "over" and an "under." This yogic effect occurs instantaneously in the enunciation. (2) *Make the surface vibrate, create tiny folds,* "as in a children's dance, combining rhythmic vowels and consonants that correspond to the interior forces of creation as to the differentiated parts of an organism." This creates a blockage capable of very specific flows, channels, membranes. "Radio and television sets are like sound walls." (3) *Crack and Graft!* "As though the circle tended on its own to open onto a future, as a function of the working forces that it shelters." Our membrane shudders through the refrain, periodically cracking open and responding to, with, "suddenly" *onto* the future. Contact. Deleuze and Guattari repeat a phrase of F. Scott Fitzgerald's from *The Crack-Up:* "This time, there are outbursts and crackings in the immanence of a rhizome.... The crack up 'happens almost without your knowing it but is realized suddenly indeed'" (199). To the extent that one can characterize the *topos* of the refrain as a "space," it is a space of continual variation as it incessantly cracks, creating itself as a (w)hole. This summarizes the operation of autopoiesis, an exocentric vision of interiority. Indeed, it is in visions that the ecstatic character of such an interiority—its reliance on being holey if not holy—announces itself with impossible, delirious utterances.

18. Deleuze and Guattari, *A Thousand Plateaus*, 73.

19. Michel Maffesoli, *The Shadow of Dionysus: A Contribution to the Sociology of the Orgy,* trans. Cindy Linse and Mary Kristina Palmquist (Albany: State University of New York Press, 1993), 44.

20. Ibid., 20.

21. About fourteen billion years ago, for example, "with the Universe having substantially calmed down from its frenetic beginnings, galaxies, stars and ultimately planets began to emerge as gravitationally bound clumps of the primordial elements." Brian Greene, *The Elegant Universe: Superstrings, Hidden Dimensions, and the Quest for the Ultimate Theory* (New York: Vintage Books, 2000), 347.

22. Maffesoli, *The Shadow of Dionysus*, 52.

23. Deleuze and Guattari, *A Thousand Plateaus*, 6.

24. http://www.diamondcutters.com/tip.html. In another context, Deleuze and Guattari write of the itinerant friendship of philosophy to thought, the need for carpenters to follow the grain and growth of wood. "I am the friend of the wood" (*What Is Philosophy?*, 9). This friendship should not be confused with any notion of "communication" or "consensus" but is instead indebted to a differential agon that renders friends indiscernible. "The basic point about friendship is that the two friends are like claimant and rival (but who could tell them apart?)" (Ibid., 4).

25. Deleuze and Guattari, *A Thousand Plateaus*, 372.

26. "Today hydrodynamic experts test the stability of fluid flow by introducing a perturbation that expresses the effect of molecular disorder added to the average flow. We are not so far from the clinamen of Lucretius" (Ilya Prigogine and Isabelle Stengers, *Order out of Chaos*, 141).

27. Deleuze and Guattari, *A Thousand Plateaus*, 372.

28. Maffesoli, *The Shadow of Dionysus*, 3.

29. Deleuze and Guattari, *A Thousand Plateaus*, 373.

30. Ibid., 378.

31. Burroughs, *The Place of the Dead Roads*, 86.

32. Ibid.

33. Ibid.

34. Pierre Lévy, *Becoming Virtual*, 43

35. James Hillman, "An Essay on Pan," in *Pan and the Nightmare*, ed. Wilhelm Heinrich Roscher and James Hillman (Dallas: Spring Publications, 1988), 52.

36. Gilles Deleuze, *The Logic of Sensation*, trans. Daniel Smith (Minneapolis: University of Minnesota Press, 2003), 25.

37. Mathematician and corporeal semiotician Brian Rotman has argued convincingly that the practice of the infinite (and the infinite regress) must be thought of as a nonlinear counting. In this scenario, the concept and mathematical operations of "infinity" are linked to an ecology of flesh. Here counting becomes more and more difficult as one "approaches" infinity, with "non-euclidean" phase transitions puncturing the ongoing refrain of counting—stroke, stroke, stroke. Rotman's treatment of "sorities" points to an understanding of infinity not as "impossible"—a logical register—but as a virtual entity, a simulacrum that emerges out of the self-organizing criticality proper to the repetition enabled by wetware and software (Brian Rotman, *Ad Infinitum*, [Stanford: Stanford University Press, 1993], 161–73).

38. Deleuze, *The Logic of Sensation*, 15.

39. Ibid., 28.

40. Henri Michaux, *Darkness Moves: An Henri Michaux Anthology*, trans. David Ball (Berkeley and Los Angeles: University of California Press, 1994), 204.

41. Paul Virilio, *The Vision Machine* (Bloomington: Indiana University Press, 1994), 2.

42. Michaux, *Darkness Moves*, 197.

43. One model of such a "blocked blockage" can be found in Jacques Monod and François Jacob's work on the *lac* operon, where the "repressor" molecule is itself "repressed" by an inducer molecule. "The logic of this negation of the negation, we may add, is not dialectical. It does not result in a new statement but a reiteration of the original one, written within the structure of DNA in accordance with the genetic code. The logic of biological regulatory systems abides not by Hegelian laws, but, like the workings of computers, by the propositional algebra of George Boole" (Jacques Monod, *Chance and Necessity: An Essay on the Natural Philosophy of Modern Biology* [New York: Knopf, 1971], 76).

44. Elizabeth Grosz, "Thinking the New" in *Symploke*, 65, no. 1 and 2 (1998): 52.

45. Stuart Kauffman, *At Home in the Universe* (New York: Oxford University Press, 1995), 23.

46. Deleuze and Guattari, *A Thousand Plateaus*, 371.

47. Félix Guattari, *Chaosmosis: An Ethic-Aesthetic Paradigm*, trans. Paul Bains (Bloomington: Indiana University Press, 1995), 21.

48. Roberto Casati and Achille C. Varzi, *Holes and Other Superficialities* (Cambridge, Mass.: MIT Press, 1994), 6.

49. Deleuze and Guattari, *What Is Philosophy?*, 204–5.

50. Ibid., 204.

51. Ibid., 205.

52. Constance Penley has worked on the frontier of this slash. See especially *NASA/Trek: Popular Science and Sex in America* (New York: Verso, 1997).

1. Representing Life for a Living

1. James Watson, *The Double Helix* (New York: Signet Books, 1968), 18.

2. Here I am borrowing from Stephen Wolfram's understanding of irreducibility—"many times there is no better way of predicting a computation than to actually run it, or to simulate it in some way"—as treated in Rudy Rucker's *Mind Tools: The Five Levels of Mathematical Reality* (Boston: Houghton Mifflin, 1987), 313.

3. François Jacob, *The Logic of Life: A History of Heredity*, trans. Betty E. Spillmann (New York: Pantheon, 1973), 299.

4. Kauffman, *At Home in The Universe*, 56.

5. Ibid., 57.

6. Ibid., 77.

7. Elias Canneti, in *Crowds and Power* (trans. Carol Stewart [New York: Noonday Press, 1962]), offers remarkable insight into the different forms—crowds, packs, mobs—such multiplicities can become. See also Deleuze and Guattari's treatment of centered versus acentered networks in "Introduction: Rhizome" (*A Thousand Plateaus*, 3–25).

8. Bretton Woods, the agreement ending the orientation of the world currency markets to the gold standard, would mark a comparable dislocation of economic value. For a discussion of the loss of reference associated with this shift, see Brian Rotman's *Signifying Nothing* (Stanford, Calif.: Stanford University Press, 1993), 87–107.

9. My argument that scientists have recently begun interfacing with organisms as informational constructs, of course, does not rule out the possibility and the probability that organisms have always already been, in part, such constructs. My point is, rather, that such an interface has only been made visible and actual via the rhetorics and practices of late-twentieth-century biology. These rhetorics and practices, therefore, disciplined and comported organisms in a way that made the informatic character of life available to the scientific register.

10. See for, example, the description of life as a "duplex" system by alife researcher Fred Cohen. While Cohen's insight concerning computer viruses treats artificial life as an ecological event—"A recipe for interactions capable of cooking up the mechanics of life required the setting that allowed its potential to be fulfilled"—it also thoroughly dislocates the emergence of life, distributing it to the (at least) parallel processes of program and instantiation, instruction and construction (Steven Levy, *Artificial Life: The Quest for a New Creation* [New York: Pantheon Books, 1992], 312). On the conflation of "construction" and "instruction," see my *On Beyond Living: Rhetorical Transformations of the Life*

Sciences (Stanford, Calif.: Stanford University Press, 1997).

11. Computer scientist Pierre Lévy characterizes the positively affective character of this excited uncertainty: "it is the vertiginous sensation of plunging into the communal brain that explains the current enthusiasm for the Internet" (*Becoming Virtual*, 145). Crucial to this analysis is the notion that something besides representation fuels the encounter with the machine—a practiced vertigo.

12. Ibid., 29.

13. Charles Sanders Peirce, *Collected Papers, Volume 2* (Cambridge, Mass.: Belknap Press of the Harvard University Press, 1932), 270.

14. An iteration of the most famous example of the syllogism will, I hope, help explicate Peirce's discussion.

All men are mortal. (Major Premise)

Socrates is a man. (Minor Premise)

Therefore, Socrates is mortal. (Conclusion)

The validity of the conclusion in syllogistic reasoning depends upon the validity of the premises. In this instance, the major premise emerges out of the persistent and habitual encounter with death, while the minor premise becomes debatable in the light of Socrates's behavior as a "gadfly."

15. Peirce, *Collected Papers*, 643.

16. Chris Langton, "Artificial Life," in *Artificial Life* (Redwood City, Calif.: Addison-Wesley, 1989), 33.

17. Peirce, *Collected Papers*, 642.

18. Langton, "Artificial Life," 2.

19. The distributed computing solution Seti@home (http://setiathome.ssl.berkeley.edu) allows users to become part of a network that trolls data for alien signals, right from your computer desktop. The problem of authentication is articulated in the software license agreement of the program, where a strange promise is acceded to: The impossible promise *not to become alarmed*. "I understand that strong signals will occasionally be detected and displayed on the screensaver during the course of data analysis. I will not get alarmed and call the press when such signals appear."

20. Langton, "Artificial Life," 24.

21. Ibid., 23.

22. Lynn Margulis, *Early Life* (Boston: Science Books International, 1982), 85.

23. Even within the nucleus of eukaryotes, there are differentials of blockage: "Chromatin is a complex of DNA and protein diffused throughout the cell nucleus most of the time. As a eukaryotic cell prepares to divide, the chromatin condenses into rod-shaped bodies—the chromosomes—and in many the nucleolus disappears" (Margulis, *Early Life*, 3).

24. Deleuze and Guattari have written of the orchid's "imaging" of the wasp as an example of a rhizomatic relation. The wasp, they argue, is an element in the reproductive system of the orchid, and as such is *part of the orchid's morphology*, its becoming-wasp, the wasp's becoming-orchid. "The orchid deterritorializes by forming an image, a tracing of the wasp; but the wasp reterritorializes on that image. The wasp is nevertheless deterritorialized, becoming a piece in the orchid's reproductive apparatus. But it reterritorializes the orchid by transporting its pollen. Wasp and orchid, as heterogeneous elements, form a rhizome" (*A Thousand Plateaus*, 10).

Darwin, by contrast, was convinced that insects must, as autonomous organisms, derive some material benefit from the pollination practice, such as nectar, while Sprengel posited the existence of what he called "Sheinsaftblumen," "or sham-nectar producers; he believes that these plants exist by an organized system of deception, for he well knew that the visits of insects were indispensable for their fertilization. But when we reflect on the incalculable number of the pollen-masses attached to their proboscides, that the same insects visit a large number of flowers, we can hardly believe in so gigantic an imposture" (Charles Darwin, *Various Contrivances by which British and Foreign Orchids Are Fertilized by Insects*, (1862; reprint, *DARWIN Multimedia CD-ROM—The Collective Works of Charles Darwin on CD-ROM* [n.p.: Lightbinders, Inc., 1997]).

25. Lévy, *Becoming Virtual*, p. 171.

26. Grosz, 38–55.

27. Compare, for example, the status of alife organisms with other boundary-troubling "life" forms. A virus's status as a life form—is it alive or isn't it?—has no effect on its ability to propagate. By contrast, the success of alife organisms is tied to the replicability, if not description, of the intensely affective responses to alife creatures. These affects are replicated in part by rhetorical softwares that comport alife *as* life and that narrate the liveliness of the creatures to ourselves or others. These narratives render us as what Simon Schaffer and Stephen Shapin might characterize as virtual witnesses to actualized life.

28. Consider, for example, alife researcher Chris Langton's close encounter of a silicon kind, as told by new media prophet Kevin Kelley: "Langton remembered working alone late one night and suddenly feeling the presence of someone, something alive in the room, staring at him. He looked up and on the screen of Life he saw an amazing pattern of self-replicating cells. A few minutes later he felt the presence again. He looked up again and saw that the pattern had died. He suddenly felt that the pattern had been alive—alive and as real as mold on an agar plate—but on a computer screen instead. The bombastic idea that perhaps a computer program could capture life sprouted in Langton's mind" (Kevin Kelly, *Out of Control* [New York: Addison-Wesley, 1994], 344; also available on-line at http://www.kk.org/ outofcontrol.index.html. Note, of course, the creepy vitality of ideas of life that themselves "sprout." See also Pier Luigi Luisi's "Defining the Transition to Life" in *Thinking About Biology*, ed. Francesco Varela and Wilfred Stein (Redwood City: Addison-Wesley, 1992), 35. Luisi writes of the affective charge carried by the claim for the synthesis of vitality: "the self-replicating bounded structures . . . should be considered as minimal synthetic life. Such a statement may possess an unappealing flavor, but I believe one should not be afraid of it. This feeling of unappealingness probably arises for psychological reasons, but it should not cloud the scientific issue."

29. Charles Darwin, *The Origin of Species* (New York: Modern Library: 1962), 69.

30. Ibid., 70.

31. Richard Dawkins defines his notion of the extended phenotype, where gene action is not confined to the interior of sovereign bodies, in terms of the following "central theorem": "An animal's behavior tends to maximize the survival of the genes 'for' that behavior, whether or not those genes happen to be in the body of the particular animal performing it" (*The Extended Phenotype* [Oxford: Oxford University Press, 1982], 233). In the case of alife creatures, then, the behavior in question is representational "life," and the maximization of the survival and propagation of this behavior is carried out, actualized, by humans. And just as the reproduction of flowering plants depend on the "imaging" of insect pollinators, so too does the actualization of alife depend upon an imaging of vitality for its human propagators. Dawkins discusses a fascinating example of annelid worms that simulated, as a group, anemones. My example here takes the notion one step further, where the very life of the organism is itself simulated as a seductive tactic.

32. John Von Neumann, *Theory of Self-Reproducing Automata*, edited and completed by Arthur W. Burks (Urbana: University of Illinois Press, 1966), 34. Remarkably, this treatise was not produced solely by von Neumann himself. Arthur Burks—who is also the editor of several volumes of Charles Sanders Peirce's selected papers—worked with the wire recorders that had preserved von Neumann's speech and rendered it into text.

33. Philip K. Dick, *Game Players of Titan* (New York: Vintage, 1992). 6.

34. Ibid., 116.

35. Ibid.

36. Philip K. Dick, *Radio Free Albemuth* (New York: Vintage, 1985), 10.

37. Ibid., 110–11.

38. Numerous burial services—both of the electronic and earthly kind—have been offered worldwide for artificial life pets. See Caroline Wyatt, "Despatches" BBC News, 20 November 1997 (http://www.cnn.com/ WORLD/9801/18/tamagotchi).

39. Some scholars argue that it was indeed the tombstone that first provided the ecology for the emergence of writing itself, an old but fundamentally erratic technology of immortality.

40. Humberto Maturana and Francisco Varela, "Autopoiesis: The Organization of the Living," in *Autopoiesis and Cognition: The Realization of the Living* (Dordrecht, Holland: D. Reidel, 1980), 73.

41. Ibid., 78–79.

42. Ibid., 102.

43. Ibid., 82.

44. Ibid., 73.

45. Ibid., 118.

46. Instructive in this regard is the ethics of Emmanuel Levinas, for whom the autonomous "subject" is constituted as the continual encounter with alterity, an encounter that renders it topologically not as a unity but as a Klein bottle: "Subjectivity realizes these impossible exigencies—the astonishing feat of containing more than it is possible to contain" (Emmanuel Levinas, *Totality and Infinity: An Essay on Exteriority*, trans. A. Lingis [Pittsburgh: Duquesne University Press, 1969], 27).

47. Maturana and Varela, *Autopoiesis*, 109.

48. Donna Haraway, *Modest_Witness@ Second_Millennium.FemaleMan_Meets_OncoMouse* (New York: Routledge, 1997), 176.

49. Haraway, *Modest_Witness*, 186.

50. Indeed, if life has the attribute of "autonomy," it has an equal measure of deterritorialization. See Nell Boyce, "Taco Trouble." *The New Scientist* 7 (October 2000) (http://www.newscientist.com/gm/gm.jsp?id=22590800).

2. Simflesh, Simbones

1. "Perhaps one of the most important characteristics of the rhizome is that it always has multiple entryways . . . the burrow is an animal rhizome, and sometimes maintains a clear distinction between the line of flight as passageway and storage or living strata (cf. The muskrat)" (Deleuze and Guattari, *A Thousand Plateaus*, 12).

2. Polymerase chain reaction: a technique for amplifying small quantities of DNA into large numbers of copies. See Kary Mullis, *Dancing Naked in the Mind Field* (New York: Vintage, 1998).

3. J. D. Bernal, "Definitions of Life," *New Scientist* 23 (1967): 34.

4. A somewhat different but no less ecstatic version of this complicity animates Jeremy Narby's suggestive but hermeneutically hamstrung book *The Cosmic Serpent: DNA and the Origins of Knowledge* (New York: Putnam, 1998).

5. Deleuze and Guattari, *A Thousand Plateaus*, 266.

6. An example of such a strata that is composed of more than nature or culture is described in Richard Preston's recent nonfiction horror bestseller, *The Hot Zone* (New York: Anchor, 1994). Writing of a highway in Kenya, Preston locates it as an ecology for HIV: "The road was once a dirt track that wandered through the heart of Africa, almost impossible to traverse along its complete length. Long sections of it were paved in the nineteen-seventies, and the trucks began rolling through and soon afterward the AIDS virus appeared in towns along the highway"(376). While marking the idiom of horror within which this text is inscribed—the book ends with a safari to find a virus in Kenya—I also want to highlight its insistence on the link between the emergence of "hot" viruses and that cyborg monster, globalized capital. In effect, *The Hot Zone* argues that transnational infrastructures—and not simply "the rainforest," "Africa," or "Nature"—are the petri dish and vector of such lifeforms. As such, these viruses are neither "natural" nor inevitable; they emerge from the mixture of asphalt, speed, and money that compose the contemporary earth.

7. Deleuze and Guattari, *A Thousand Plateaus*, 270.

8. Artificial life emerges out of several domains, but the hegemonic desire is for an understanding of the universal character of life. On this account, biology has thus far been hamstrung by its dependence on only one form of life, carbon-based life. Thus alife, so the story goes, gives researchers access to more than life as we know it. It allows our gaze to fall on "life as it could be." See the International Society for Artificial Life at http://www.alife.org/.

9. Stefan Helmreich, an anthropologist who studied the emergence of alife and complexity as scientific objects, has produced an extraordinary account of these practices. See his *Silicon Second Nature* (Berkeley and Los Angeles: University of California Press, 1998).

10. David Gelernter, *Mirror Worlds: The Day Software Puts the Universe in a Shoebox... How It Will Happen and What It Will Mean* (New York: Oxford University Press, 1991), 52.

11. Ibid., 1.

12. See also Deleuze and Guattari's treatment of smooth versus striated space in *A Thousand Plateaus* (474–500).

13. Deleuze and Guattari, *A Thousand Plateaus*, 372.

14. Rotman's mousegrams live at http://www.accad.ohio-state.edu/~brotman/mousegrams.html.

15. *Science* 256 (8 May 1992): 911.

16. Ibid.

17. "First, the posthuman view privileges informational pattern over material instantiation" (N. Katherine Hayles, *How We Became Posthuman* [Chicago: University of Chicago Press, 1999], 2).

18. Ibid.

19. Walter Gilbert, "Towards a paradigm shift in biology," *Nature* 349 (10 January 1991): 99.

20. Michael Bremer, *SimLife: The Genetic Playground* (Orinda, Calif.: Maxis Corporation, 1992), 4.

21. "Use of computer technology often has unpleasant side effects, some of which are strong, negative emotional states. These negative emotional states can affect not only the interaction with the computer, but productivity, learning, creativity, and overall well-being." See Jonathan Klein, Youngme Moon, and Rosalind W. Picard, "This Computer Responds to User Frustration," Vision and Modeling Group Technical Report 502, MIT Media Laboratory, 17 June 1999. Available at http://vismod.www.media.mit.edu/tech-reports/TR-502/index.html.

22. Bremer, *SimLife*, 82.

23. Ibid., 12.

24. William S. Burroughs, *The Adding Machine* (New York: Seaver Books, 1985), 18.

25. See, for example David Heibeler's "Implications of Creation," a working paper of the Santa Fe Institute available at http://www.santafe.edu/sfi/publications/wpabstract/199305025.

26. Karl Sigmund, *Games of Life: Explorations in Ecology, Evolution, and Behaviour* (New York: Oxford University Press, 1993), 10.

27. Steven Levy, *Artificial Life: Quest for a New Creation* (New York: Pantheon Books, 1993), 95.

28. Sigmund, *Games of Life*, 10.

29. This scattering of the very possibility of representing organisms as objects—rather than flickers, events—will be taken up in more detail in a discussion of cryonics.

30. Frank Miele, *The (Im)moral Animal: A Quick & Dirty Guide to Evolutionary Psychology & the Nature of*

Human Nature. Available at http://www.skeptic.com/04.1.miele-immoral.html.

3. Disciplined by the Future

1. This message, along with the epigraph to this chapter and many messages that follow, can be found at the CryoNet archive located at http://keithlynch.net/cryonet/. The archive is searchable by keyword and message number, so the remaining references to the archive will include the message number. This is also a good place to find links to other resources on cryonics.

2. As befits their practice of *tinkering with the future*, cryonic subscribers have appropriated the de-territorializing capacities of the refrigerator—Yum! Raspberries in December in New York!—transforming it into a component of a machine that freezes bodies and not just vegetables out of season.

3. "The elementary unity of language—the statement is the order word." Deleuze and Guattari, *A Thousand Plateaus*, 76.

4. Indeed, most people fund cryonics with life insurance! From the FAQ at the CryoNet archive (http://keithlynch.net/cryonet/): "The person who makes the cryonics arrangements pays for suspension, usually with life insurance. Some life insurance companies refuse to accept a cryonics organization as the beneficiary. Check with your insurance agent, or check with a cryonics organization for a list of cooperative companies."

5. Indeed, the odds of cryonics success—here interpreted as becoming-popular—are themselves in a futures market, "the Idea Futures Market of Alberta Canada" located at http://groups.google.com/groups?q=idea+futures+market&hl=en&lr=&ie=UTH-8&selm=morgan-3110941611590001%40hathi.arc.ab.ca&rnum=5. Current odds that cryonics will catch on—that at least 10,000 patients will be frozen by 31 December 2000—are 13–19 percent. These odds are suitably similar to the odds of nanotechnological success (a Stewart platform) which run at 11–13 percent, and time travel, which are 8–14 percent. Its relation to the odds on Quebec separating from Canada—currently nearly identical to cryonics' odds at 13–14 percent—is less clear. There is also much discussion of a "secondary" market in cryonics, whereby cryonicists could sell or buy memberships in a cryonics organization independently of the organization. Presumably, an investor with great confidence in the feasibility of cryonics could buy several memberships to be resold after demand responds to the cryonic success, making what's known in all markets as a "killing." From Newsgroups: sci.cryonics: Message-ID: Cryo.786537010.20736@skyler.arc.ab.ca Organization: Alberta Research Council. Sunday, 4 December 1994.

6. Alcor, *Cryonics: Reaching for Tomorrow*, 1996 (http://www.alcor.org/CRFTnew/crft05.htm).

7. Franz Kafka, *The Trial* (New York: Pantheon Books, 1998).

8. Emmanuel Levinas, *Totality and Infinity: An Essay on Exteriority* (Pittsburgh: Duquesne University, 1969).

9. Robert Ettinger, *The Prospect of Immortality* (Garden City, N.Y.: Doubleday, 1964).

10. Deleuze and Guattari, *A Thousand Plateaus*, 86.

11. Levinas, *Totality and Infinity*, 27.

12. This FAQ, as well as all others pertaining to Alcor, is available at http://www.alcor.org/FAQs.

13. This FAQ is available at http://www.faqs.org/faqs/cryonics-faq/part1/.

14. One can only speculate on the relationship between this rumor/truth and the fact that Disney was a great twentieth-century animator.

15. Dewars are the storage containers used in cryonics.

16. Hans Moravec, *Mind Children: The Future of Robot and Human Intelligence* (Cambridge, Mass.: Harvard University Press, 1988), 4.

17. According to Stephen Manes and Paul Andrews in *Gates: How Microsoft's Mogul Reinvented an Industry—And Made Himself the Richest Man in America* (New York: Doubleday, 1993), Gates "began talking about his pet topic, biotechnology: how genetic code was similar to binary code, how in the future humans would be able to be downloaded onto chips, the human spirit would be burned into silicon, and silicon-based life forms would replace those based on mere carbon" (456).

18. Ralph Merkle, "Cryonics, Cryptography, and Maximum Likelihood Estimation" (1994) (http://www.merkle.com/merkleDir/cryptoCryo.html).

19. Ibid.

20. Philip K. Dick's *A Scanner Darkly* (N.Y.: DAW Books, 1984) features precisely such a disturbing scenario. "Fred," a narcotics agent, finds that it is his own undercover persona, Bob Arctor, that he is investigating.

21. This becoming could, of course, be rendered algorithmically in terms of fractal geometry, whose purview is precisely such "self-similar" structures. But such a fractal mapping of subjectivity would provoke the question: Mapped for what or whom? The criterion of mapping and "freezing" the frame of a fractal arrests the multiplicity of its mapping, and the arrest is carried out in terms of the mapping subject, even if that subject is only "chance." Of a similar problem, that of the "fractal" mapping of a shoreline, Brian Massumi writes: "Every moment in life is a step in a random walk. Uncannily familiar as the shore may seem, looking back reveals no Eden of interiority and self-similarity" (*A User's Guide to Capitalism and Schizophrenia: Deviations* *from Deleuze and Guattari* [Cambridge, Mass.: MIT Press, 1992], 22).

22. Merkle, n.p.

23. Robert F. Nelson, *We Froze the First Man: The Startling True Story of the First Great Step toward Human Immortality* (New York: Dell Publishing, 1968), 56.

24. For other formulations of this notion of "Eternal Life," see Frank Tipler, *The Physics of Immortality: Modern Cosmology, God, and the Resurrection of the Dead* (New York: Anchor Books, 1995), 11. Note that Tipler's scheme of immortality relies on a practice of obliterating difference, as the universe itself becomes nothing but a vessel for the endless replication of human identities, universal consumers whose only goal is high-fidelity repetition. By contrast, cryonic bodies are differential by design, a telos whose actualization requires an outburst of radical difference: the future. Thus the cryonic body, as a virtual body, a body becoming code, is also a body without organs, Deleuze and Guattari's characterization of a corporeal ecology that eludes transcendence and subjectification through practices of territorialization and deterritorialization. See Deleuze and Guattari's how-to manual for the production of these states, "November 28, 1947: How Do You Make Yourself a Body Without Organs?" *A Thousand Plateaus*, 149–66.

25. Rotman, *Becoming Beside Oneself* (http://www.stanford.edu/dept/HPS/WritingScience/etexts/Rotman/Becoming.html).

26. Rotman, *Signifying Nothing*, 12.

27. Ibid., 19.

28. Numa Financial Systems, "The Derivatives FAQ" (1995) (http://www.numa.com/ref/faq.htm#no01).

29. House Committee on Energy and Commerce, *Derivative Financial Markets: Hearings before the Subcommittee on Telecommunications and Finance*, 103rd Cong., 1994, 2.

30. House Committee, 1994, Part 2, 91.

31. House Committee, 1994, Part 2, 1.

32. Like the computer program with which it is linked, the derivative must be "run" in order to determine its results.

33. Fredric Jameson, *Postmodernism, or, The Cultural Logic of Late Capitalism* (Durham, N.C.: Duke University Press, 1991), 411.

34. House Committee, 1994, Part 2, 105–6.

35. Ettinger, 1964, xi.

36. Consider the case of a recent cryonics patient. Having decided that his terminal illness no longer allowed him to endure life, he chose euthanasia for relief. Normally, this precision in the decision of death would be thought to aid cryonics, for cryonics needs as much certainty about death as it can get to enable the tech team time to begin their treatment of the body in the minimum amount of time after death. In this instance, however, even the fulfillment of the contract of euthanasia was thwarted, as a virus had infected many of the members of the cooldown team. The cryonics organization requested that the patient defer his death in order to allow him to be "prepared." The patient, installed with a morphine pump, decided to comply with the cryonics organization's last wish, a wish for the future that disciplined his body—he obediently did not die until the cooldown team had recovered from the virus.

Note also that in these cases the body is often prepared—with a varied mixture of drugs—for a day before actual death, inscribing the body with drugs in anticipation of a future death.

4. "Give Me a Body, Then"

1. Gilles Deleuze, *Cinema 2: The Time-Image*, trans. Hugh Tomlinson and Robert Galeta (Minneapolis: University of Minnesota Press, 1989), 215.

2. Ibid., 272.

3. Gilles Deleuze, *Cinema 1: The Movement-Image*, trans. Hugh Tomlinson and Barbara Habberjam (Minneapolis: University of Minnesota Press, 1986), 1.

4. Deleuze, *Cinema 2*, 277.

5. Ibid., 269.

6. Ibid., 215.

7. Ibid., 189.

8. Ibid., 212.

9. Ibid., 280.

10. Walter Gilbert, "Towards a Paradigm Shift in Biology," *Nature* 349 (10 January 1991): 99.

11. Merkle, n. p.

12. Irreversibility is quite literally a non sequitur within classical models of thermodynamics, where time and its ongoing action are assumed to be mere illusion and sequences of events can be run backwards and forwards in time. In this sense Merkle's mourning indicates an outburst of biological reality: "We have discovered that far from being an illusion, irreversibility plays an essential role in nature and lies at the origin of most processes of self organization" (Prigogine and Stengers, 9).

13. Stuart Kauffman finds himself astonished that such conditions of possibility for a biosphere (that largest unit of and apparent obligatory passage point for life) cannot be articulated in advance. Writing of his conclusion that life cannot be scientifically compared to an algorithmic procedure, a procedure of instantiating a sequence, Kauffman writes: "I will argue that we cannot prestate some biological analogue of the input data, nor is there some biological analogue of the program governing the unfolding of a biosphere" (*Investigations* [New York: Oxford University Press, 2000], 123). One might ask why any good empiricist would ever think it *were* possible to prestate such a source of continuous novelty, and my preliminary answer would be: by stealthing the complexity. For an account of how such complexity was productively hidden in the rise of molecular biology, see Richard Doyle, *On Beyond Living: Rhetorical Transformations of the Life Sciences* (Stanford: Stanford University Press, 1997), especially chapter 5.

14. Deleuze, "Mediators," in *Incorporations*, ed. Jonathan Crary and Sanford Kwinter (New York: Zone, 1992), 283.

15. "It is here that we may speak the most precisely of the crystal-image: the coalescence of an actual image and *its* virtual image, the indiscernibility of two distinct images" Deleuze, *Cinema 2*, 127.

16. Ibid., 89.

5. "Remains to Be Seen"

1. Alcor explains the move as follows: "By 1993 the need for office space alone was eating up room at Alcor's Riverside facility, but the constant addition of new patients and their storage capsules left staff members feeling as though the building would burst at the seams any day. Then too, there was the Natural Disaster problem; considering the regular stream of earthquakes, brushfires, and mudslides (not to mention the smog, crime, and riots) reported from California it seemed ridiculous to maintain a long-term storage facility there. Finally, the Riverside City Council, vehemently against animal experimentation, shut down almost all of Alcor's research. Alcor Headquarters had to move" (http://www.alcor.org/historyb.html).

2. *Prospect* was first published privately by Ettinger in 1962, and picked up by Doubleday in 1964 (Garden City, N.Y.: Doubleday, 1964).

3. R. C. W. Ettinger, *Man into Superman* (New York: St. Martin's Press, 1972), available at http://www.cryonics.org/1chapter10.html#Beyond.

4. All Usenet group postings can be found through Google Groups. Sci.nano can be found at http://groups.google.com/groups?hl=en&group=sci.nanotech.

5. Neal Stephenson, *The Diamond Age; or, A Young Lady's Illustrated Primer.* (New York: Bantam Books, 1995).

6. Of the necessary substrates for living systems and selfhood, Ettinger writes "There is no substitute for

carbon . . . and very possibly, if you want a self circuit, there is no substitute for meat." CryoNet, Message 8061. For more on the self circuit, see my discussion that follows.

7. This "collective" memory would function only in a disciplined community of interpreters—even electron micrographs require a community of interpreters to relate them to the status of a body. In Merkle's account, this collective memory could *emerge* from a swarm of nanotechnological entities that repeat all possible pathways of evolution in a frenetic search engine for what he calls a "warmer" rather than "colder" body of revival.

8. Phillip Gourevitch, "Postscript: Recalling a Hard Life's Work," *The New Yorker* (28 December 1998 and 4 January 1999): 42.

9. I try to remember William Burroughs's missives from the Interzone. "*Naked Lunch* is a blueprint, a How-To book . . . How to extend levels of experience by opening the door at the end of a long hall . . . Doors that only open in *silence* . . . *Naked Lunch* demands Silence from The Reader. Otherwise, he is taking his own pulse." *Naked Lunch* (New York: Grove Press, 1959), 224.

10. Robert Ettinger, telephone interview by the author, tape recording.

11. In *Man into Superman* (available at http://www.cryonics.org/), Ettinger notes that cryonics, too, seeks to manage the moment of death. "In fact, the Cryonics Society of Michigan in 1970—following a suggestion a couple of years earlier of Dr. M. Coleman Harris—began investigating the legal feasibility of 'mercy freezing' or freezing a terminally ill patient before clinical death. The three main advantages are obvious: (1) the patient will be less deteriorated if the presently incurable illness is not allowed to go its full course, and therefore fully successful revival will be more probable; (2) even more important, perhaps, the freezing will take place at a selected time and place, under optimum conditions, whereas ordinarily it is extremely difficult to make an accurate prediction of the date of death, so there is usually a delay after death before the team can reach the patient; (3) suffering and expense will be reduced."

12. Mike Darwin, a prominent cryonics researcher now with Cryocare, ran the Institute for Advanced Biological Studies, in Indianapolis, Indiana in the late 1970s. According to Alcor's online history, "by 1981 Mike decided that the Midwest was no place for cryonics to grow, and moved to California" (http://www.alcor.org/historyb.html).

13. *Tetsuo*, dir. Shinya Tsukamoto, Fox Lorber Home Video, 1998. Videocassette.

14. *Oxford English Dictionary*, 2d ed., s.v. "amalgam."

15. Ibid.

16. There are a diverse set of concepts and practices associated with uploading, but all entail the replication of human identity and/or "consciousness" in a computer environment. This concept was popularly articulated by computer scientist Hans Moravec in his 1988 book, *Mind Children*, but has many literary, scientific, and philosophical antecedents. For analysis on the rhetorical and affective practices of uploading, see chapter 6 of *Mind Children*. For speculative articulation of uploading's effects, see Robin Hanson's "If Uploads Come First: The Crack of a Future Dawn" (http://hanson.gmu.edu/uploads.html).

17. Marvin Minsky, *Society of Mind* (New York: Voyager, 1994).

18. Burroughs, *The Adding Machine*, 135.

19. Spookier still, historian of science Michael Fortun and physicist Herb Bernstein have played a veritable cat's cradle with these entanglements, for whom they present a case study in the contingency and power of ongoing research: the entanglement of technoscience with the future. "In a sense, all scientists are emissaries to a strange and creative place in the world that might be called the future of everyday life" (*Muddling Through: Pursuing Science in the Twenty-first Century* ([Washington, D.C.: Counterpoint, 1998], 232). Ironically, Fortun and Bernstein's analysis highlights the role such entanglements might play in cryptography, where they offer the promise of breaking all public key encryption technologies, perhaps aiding Merkle's plans for future decoding. At the same time, this specter of quantum computing also promises to produce the holy grail of cryptography and a fearful event for cryonics—an unbreakable code.

20. Ettinger, *Man into Superman* (http://www.cryonics.org/chapter5_1.html).

6. Uploading Anticipation, Becoming Silicon

1. Donna Haraway, *Simians, Cyborgs, and Women: The Reinvention of Nature* (New York: Routledge, 1991).

2. Elizabeth A. Wilson, *Neural Geographies: Feminism and the Microstructure of Cognition* (New York: Routledge, 1998), 162.

3. Gilles Deleuze, *Foucault*, trans. Seán Hand (Minneapolis: University of Minnesota Press, 1988), 131.

4. Tipler, *The Physics of Immortality*.

5. Scott Bukatman, *Terminal Identity: The Virtual Subject in Postmodern Science Fiction* (Durham: Duke University Press, 1993), 16.

6. Ibid., 20.

7. Allucquere Rosanne Stone, "Will the Real Body Please Stand Up?" in *Cyberspace: First Steps*, ed. Michael Benedikt (Cambridge, Mass.: MIT Press, 1991), 99.

8. Michel Foucault, *Technologies of the Self: A Seminar with Michel Foucault* (Amherst: University of Massachusetts Press, 1988), 18.

9. Ibid., 27.

10. Ibid., 47.

11. Ibid.

12. Ibid.

13. Ibid., 48.

14. Philip K. Dick, *Do Androids Dream of Electric Sheep?* (New York: Ballantine Books), 1982.

15. William S. Burroughs, "Immortality," in *The Adding Machine*, 132.

16. CryoNet online discussion group, message 5310.

17. Gilles Deleuze, "Spinoza and Us," in *Incorporations*, ed. Jonathan Crary and Sanford Kwinter (New York: Zone Books, 1992), 281.

18. Rotman, *Signifying Nothing*, 93.

19. William Gibson, *Neuromancer* (West Bloomfield, Mich.: Phantasia Press, 1986). The literary history of uploading, of course, extends well beyond Gibson. Thomas Pynchon wrote of "electrofreaks" in *Gravity's Rainbow* (New York: Viking, 1973). Here Pop gives Slothrop a lecture on the danger of the electronic soul:"—Listen Tyrone, you don't know how dangerous that stuff is. Suppose someday you just plug in and go away and never come back? Eh?"—Ho, ho!... You're such an old fuddy-duddy...Maybe there *is* a machine to take us away, take us completely, suck us out through the electrodes of the skull 'n' into the Machine and live there forever with all the other souls it's got stored there... Dope never gave you immortality." (699). See also Philip K. Dick, who wrote of a slightly more eroticized uploading, an amplified precursor to Internet sex, in *Flow my Tears, the Policeman Said.* (New York: Vintage, 1993).

20. Gibson, *Neuromancer*, 106.

21. Ibid.

22. Ibid., 105–6.

23. Ken Karakotsios and Justin V. McCormick. *SimLife: The Genetic Playground* (Orinda, Calif.: Maxis, 1992).

24. Merlin Donald, *Origins of the Modern Mind: Three Stages in the Evolution of Culture and Cognition* (Cambridge, Mass.: Harvard University Press, 1991), 356.

25. Of course, the demographic profile of uploading enthusiasts does bear the stamp of such constitutive categories as race. My point here is that in the operation of the discourse circuit by which self "fashioning" occurs within uploading, race and gender are not simply unmarked categories: they are usurped categories, subsumed by the irreducible risk of the future, a future in which race is very much *at stake*.

26. Robin Hanson, *If Uploads Come First*, n.p.

27. K. Eric Drexler, *Engines of Creation* (Garden City, N.Y.: Anchor Books/Doubleday, 1990, 1986). Also available at http://www.foresight.org/EOC/index.html.

28. Drexler seeks to increase the quality of this anticipation in the present through new models of hypertext publishing, rhetorical tools that will help us to avoid the mistakes that would haunt or foil any nanotechnological future. "We face many other big, messy problems where discussion is up against one or more of the above limits. Examples include acid rain, ozone depletion, nuclear winter, genetic engineering, nanotechnology, economic policy, and military strategy. Many of these issues are cross-disciplinary, involving chemistry, physics, biology, ecology, economics, political science, and so forth. All are complex, involving economic systems, ecosystems, multiple technologies, international politics, and so forth. All are subjects of contention. In all of them, an improved chance of avoiding major mistakes could be of enormous value" (http://www.foresight.org/EOC/index.html).

29. Susan Oyama, "The Accidental Chordate: Contingency in Developmental Systems," *South Atlantic Quarterly* 94 (1995): 511.

30. CryoNet, message 5310. See also Rudy Rucker, *Mind Tools: The Five Levels of Mathematical Reality* (Boston: Houghton Mifflin, 1987).

31. Hanson, *If Uploads Come First*, n.p.

32. Moravec, 114.

33. Hanson, n.p.

34. CryoNet, message 5304.

35. Ibid.

36. Deleuze and Guattari, *A Thousand Plateaus*, 59–60.

37. Take a strip of paper. Fold it in half. Leave it undisturbed for a few days. Notice the dust and bacteria that makes up the ecology of the fold. Now take the strip and twist it. What ecologies are enabled by this new surface?

38. Jean Baudrillard, *Simulacra and Simulation*, trans. Sheila Faria Glaser (Ann Arbor: University of Michigan Press, 1994), 149.

39. *Videodrome*, dir. David Cronenberg, Universal Studios, 1982. Videocassette.

40. Alluquere Rosanne Stone, "Will the Real Body Please Stand Up? Boundary Stories About Visual Cultures," in *Cyberspace: First Steps*, ed. Micahel Benedikt (Cambridge, Mass.: MIT Press, 1991), 113.

7. Dot Coma

1. Doyle, *On Beyond Living*.

2. For a remarkable account of this new biological concept of the self, see Dorion Sagan's "Metametazoa: Biology and Multiplicity," where he notes that "the zoological 'I' is open to radical revision" (in *Incorporations*, ed. Jonathan Crary and Sanford Kwinter [New York: Zone Books, 1992], 379). See also Margulis, Sagan, and Morrison, *Slanted Truths*.

3. "Gary Dockery, the former Walden police officer who regained the power of speech last year after 7 1/2 years in a comalike state, died Tuesday at a Signal Mountain nursing home." See "Police Officer's Memorial," http://www.behindthebadge.net/pmemorial/pmem_d.html#D.

4. http://iris.npr.org/plweb-cgi/fastweb?getdoc+npr+npr+15699+0+wAAA+2%2F5%2F96%26And%26coma.

5. http://www.cbn.org/news/archives.asp.

6. http://www.columbia-hca.com/.

7. Ibid.

8. http://iris.npr.org/plweb-cgi/fastweb?getdoc+npr+npr+15699+0+wAAA+2%2F5%2F96%26And%26coma.

9. Steven Shapin and Simon Schaffer describe the emergence of "virtual witnessing" as the production of iteration at a distance in their account of Boyle in *Leviathan and the Air Pump: Hobbes, Boyle, and the Experimental Life* (Princeton, N.J.: Princeton University Press, 1985). Here the persuasive evidence of Boyle's air pump was replicated through writing, in the absence of Boyle or the pump but in the "presence" of the reader. In Dockery's instance, we see a virtual witnessing that operates through the absence not of the other, but of the self.

10. "A Definition of Irreversible Coma. Report of the Ad Hoc Committee of the Harvard Medical School to Examine the Definition of Brain-Death," *Journal of the American Medical Association* 205 (1968): 337–40.

11. John Cheyne, *Cases of Apoplexy and Lethargy: With Observations upon the Comatose Diseases* (London: T. Underwood, 1812).

12. While there is a continuous thread of ambiguity that runs through the genealogy of the coma, the question of medicine's relation to such ambiguous states is marked by profound discontinuities. For example, Cheyne writes of the apoplectic disorders almost entirely from the perspective of the doctor-patient relation and seldom names any other parties—such as a family—who might have an interest in this sudden eruption of disorder. See his *Cases of Apoplexy and Lethargy*.

13. http://www.pitt.edu/~cep/41–3.html.

14. Michelle Williams, "Former Policeman Continues to Speak," *The Daily Iowan*, 7 April 1996 (http://www.uiowa.edu/~dlyiowan/).

15. For a detailed analysis of how families can "arouse" such bodies, see Edward B. LeWinn, *Coma Arousal: The Family as a Team* (Garden City, N.Y.: Doubleday, 1985).

16. See the Dockery family's Web page at http://www.tib.com/dfai/.

17. Avital Ronell, *The Telephone Book: Technology—Schizophrenia—Electric Speech* (Lincoln: University of Nebraska Press, 1989).

18. Stephen King, *The Dead Zone* (New York: Viking Press, 1979) and *The Dead Zone*, dir. David Cronenberg, Paramount, 1983.

19. See, for example, Irvine Welsh's *Marabou Stork Nightmares* (New York: W. W. Norton, 1996). Here our narrator—in a coma—is incessantly interrupted by his "genetic disaster" of a family, a family that insists on retelling familial tales to the prone Roy Strang.

20. Even the Zetas—allegedly alien entities that purport to answer questions on the Web—note the integral relation between comatose patients and the masochistic project of waiting: "Comatose patients are very distressing to doctors as there is essentially nothing to be done except wait. With an infection the doctor can try various antibiotic or heat treatments, enrich the patient's diet, and perhaps even work on their psychological state to boost the immune response. But except for maintenance of the human body, there is nothing to be done for a comatose patient" (available at: http://www.zetatalk3.com/science/s43.htm).

21. Here of course I am referring not to "high" literature, but to the capacity for being affected (a getting-high) associated with the very possibility of reading or of viewing with intensity. This capacity emerges not out of an act of sheer agency on the part of the reader or viewer—that would be like laughing, trembling, on purpose—but to a paradoxical agency of *possession* by assemblage. The oft-cited "suspension of disbelief" refers to an action *on* the reader and not just *of* the reader. One is rendered sensitive to signs, a receptivity to the future that entails an "undoing" or suspension of the subject in the present. Such a suspension—which need not deploy ropes or chains to provoke that most masochistic of practices, waiting—summons a limbo agency detailed most coherently by Deleuze in his analysis of Leopold von Sacher-Masoch novels. See Deleuze's *Coldness and Cruelty*, 70–72. See also Deleuze and Guattari's *A Thousand Plateaus:* "Learning to undo things, and to undo oneself, is proper to the war machine: the "not-doing" of the warrior, the undoing of the subject" (400).

22. Catherine Clément, *Syncope: The Philosophy of Rapture*, trans. Sally O'Driscoll and Deirdre M. Mahoney (Minneapolis: University of Minnesota Press, 1994).

23. Gilles Deleuze discusses this imaging of time's discontinuity or complexity in *Cinema 2: The Time-Image*, 215.

24. Deleuze and Guattari, *Anti-Oedipus*, 37.

25. Clearly, this silence is not an absence; it is the "interstice" out of which contemporary capital is woven. Rather than a lack of knowledge, this unbearable contingency is often itself *desired*. Brad Wieners, a reviewer for *Wired*, suggests that readers look to the architecture of Rem Koolhas to "inspire" them in their own professions so that their work is "something you want unknowable so that you may continue to make discoveries." (*Wired*, June 1997, 160).

26. "The expanding formalism of scientific visualization with its accompanying hermeneutic demands produces, in short, what might be termed a referential panic." See Karen Newman, *Fetal Positions* (Stanford, Calif.: Stanford University Press, 1996), 110.

27. In *Recalled to Life: The Story of a Coma* (New Haven: Yale University Press, 1990), Esther Goshen-Gottstein tells of the visual transformation of her husband into a patient amid the ecology or "forest" of the coma: "Dressed in a white gown with a surgical mask and dust covers for my shoes, I entered the recovery room but could hardly see Moshe among the forest of tubes and machines. There was not only the respirator to help him breathe and the balloon pump that helped his heart to beat and improved the flow of blood to the vital organs, but also innumerable tubes coming out of his body and diverse fluids dripping into his veins. Above his head, various monitors reflected his vital signs, displaying an array of ever-changing numbers. It was an awesome sight and not one that enabled me to relate in a personal way to the patient on the bed, who was my husband" (8).

28. For an excellent analysis of political and rhetorical context of the comatose mother in the 1980s, see Valerie Hartouni's "Containing Women: Reproductive Discourse in the 1980s," in *Technoculture*, ed. Constance Penley and Andrew Ross (Minneapolis: University of Minnesota Press, 1991), 27–56. More on Hartouni follows.

29. King, 62.

30. Hartouni, 30.

31. Ibid., 33.

32. Indeed, the *Oxford English Dictionary*'s etymological treatment of "medium" is organized around this "middle quality"—the quality of being between.

33. Hartouni, 37.

34. Gilles Deleuze, *Difference and Repetition* (New York: Columbia University Press, 1994), 208–9.

35. Ibid., 215.

36. Ibid.

37. Hartouni notes that only a reading that translated motherhood as "merely" biological would render the headline coherent. How could the *practice* of motherhood be carried out by anything brain dead? At the same time, Hartouni suggests that with a working understanding of motherhood that emphasized the "social activities and meanings" of that practice, "the headline itself would be virtually unintelligible" (33). More on virtual intelligibility follows.

38. This "privilege" of the cut—its crucial role in the very operation of cinema—is in fact marked by the character of Smith's second sight: his "visions" are marked *as cinematic*, as if the distance between Smith and the cinema itself imploded. See the discussion of "tactile convergence" that follows.

39. Steven Shaviro, *The Cinematic Body* (Minneapolis: University of Minnesota Press, 1993), 52.

40. Ibid., 55.

41. King, 94.

42. Indeed, this figure of the comatose body cleaved from its machinic rhizome has appeared in Usenet disputes over abortion on talk.abortion. Speaking of the fetus, on Thursday, 7 Nov 1996 23:18:14 GMT, ray@netcom.com (Ray Fischer) wrote: "it is as separate from the mother as a coma victim is from his machines" (available from www.groups.google.com).

43. King, 95.

44. Ibid., 88.

45. Ibid., 115.

46. Ibid., 110.

47. Deleuze and Guattari, *What Is Philosophy?*, 158.

48. King, 109.

49. Deleuze and Guattari, *Anti-Oedipus*, 36

50. Robin Cook, *Coma* (New York: New American Library, 1977); and, *Coma*. dir. Michael Crichton, MGM, 1978. Film. Starring Rip Torn, Michael Douglas, and Richard Doyle.

51. The National Kidney Foundation, for example, notes that "More than 50 people can be helped by one organ and tissue donor. One donor can:
- Donate kidneys to free two people from dialysis treatments needed to sustain life;
- Save the lives of patients awaiting heart, liver or lung transplants;
- Give sight to two people through the donation of corneas;
- Donate bone to help repair injured joints or to help save an arm or leg threatened by cancer or trauma;
- Save the lives of burn victims and help them heal more quickly through the donation of skin;
- Provide healthy heart valves for someone whose life is threatened by malfunctioning or diseased valves.
- Every day, 8 to 10 people die waiting for organ & tissue transplantation (http://www.kidney.org).

52. "Burns, Baby Burns," *The Simpsons*, production code 4F05, written by Ian Maxtone-Graham, dir. Jim Reardon. Originally aired in the United States 17 November 1996.

53. The rhetorical situation of the donation and request of human organs—most states have laws that require such requests from next of kin—extends also to the very treatment of transplant recipients. Patient compliance with immunosuppressant therapy is an area of intense inquiry and anticipatory surveillance by pharmaceutical organizations like Novartis (formerly Sandoz). As part of an initiative in their Transplant Learning Center, "the firm is testing methods their pharmacists can use to predict which patients are most at risk for noncompliance" (Stadtlander's Pharmacy reprint, from *Drug Topics*, 20 April 1998). In this instance, the TLC appears to take literally what would seem to be a rhetorical question that they themselves pose: "Transplant Learning Center is based on the concept of 'Who knows the patients better than themselves?'" My answer: TLC.

54. "The concept . . . refers not to a series of numbers but to strings of ideas that are reconnected over a lacuna (rather than linked together by continuation)" (Deleuze and Guattari, *What Is Philosophy?*, 161).

55. http://www.kidney.org/general/aboutNKF/od.cfm. Here the family becomes the site for the "execution" of the rhetorical algorithm of donation.

56. Families, for example, are the crucial demographic for organ donations: donor cards are indications of the desires of the deceased, desires that must be consented to by the family (http://www.organdonor.gov).

57. http://www.careproject.net/pa_declar.htm.

58. Gilles Deleuze, *Spinoza: Practical Philosophy* (San Francisco: City Lights Books, 1988), 18.

59. Grosz, "Thinking the New," 52.

60. King, 163.

61. "A Very Difficult Pregnancy," February 5, 1996, http://www.time.com/time/magazine/archives/advanced.

62. Ibid.

63. Colleen Smith, "An Unbreakable Faith," *Our Sunday Visitor*, 24 March 1996, 10–11.

8. "Take My Bone Marrow, Please"

1. "But for the dispars as an element of nomad science the relevant distinction is material-forces rather than matter-form. Here it is not exactly a question of extracting constants from variables but of placing the variables themselves in a state of continuous variation" (Deleuze and Guattari, *A Thousand Plateaus*, 369). The "variable" here is, of course, the very borders and autonomy of human bodies.

9. Wetwares

1. *Transplant Video Journal* 5 (fall 1997). Producer Stu Katz. Elm City Communications (http://www.transplant-directory.com/TranVideo.htm).

2. Arthur Kroker and Marilouise Kroker, *Hacking the Future: Stories for the Flesh-Eating 90s* (New York: St. Martin's Press, 1996), 9.

3. Ibid., 10.

4. "The earth, said he, hath a skin; and this skin hath diseases. One of these diseases, for example, is called man." Friedrich Nietzsche, *Thus Spake Zarathustra*, trans. Thomas Common (New York: The Modern Library, 1954), 143.

5. Kroker and Kroker, 11.

6. http://www.poptel.org.uk/nuj/nmd/.

7. Burroughs wrote that the cut-up was most of all a way of traversing time: "When the reader reads page ten he is flashing forward in time to page one hundred and back in time to page one." William S. Burroughs, "The Future of the Novel," in *Word Virus: The William S. Burroughs Reader*, ed. James Grauerholz and Ira Silverberg (New York: Grove Press, 1998), 272. See also Burroughs's discussion of spacetime and his deployment of the cut-up in William S. Burroughs and Allen Ginsberg, *The Yage Letters* (San Francisco: City Lights, 1988), 44–46, 59.

8. Rucker, *Mind Tools*, 290.

9. Burroughs, *The Adding Machine*, 53.

10. Burroughs, *Naked Lunch*, 17.

11. Burroughs, *The Ticket that Exploded*, 83.

12. Burroughs, *Naked Lunch*, 203.

13. The enormous complexity of actualizing DNA into proteins is the object of an early effort at distributed technoscience, where networks of personal computers and humans are clustered for massively parallel computations. This distributed or even ecological model of knowledge production shifts inquiry away from a serially algorithmic and irreducible model of living systems and uses the massively interconnected character of living systems as a computational engine. Ironically, it is DNA itself that has been the focus of early research into much massively parallel computation to solve NP complete problems such as protein folding. See Katie Pennicott "'DNA Computer' Cracks Code," *Physicsweb*, 15 March 2002 (http://physicsweb.org/article/news/6/3/11). To participate in the distributed computation of protein folding, see http://folding.stanford.edu.

14. Or indeed, to write. See, for example, Daniel Dennett, *Darwin's Dangerous Idea: Evolution and the Meanings of Life* (New York: Simon and Schuster, 1996), especially 181–85.

15. C. D. B. Bryan, *Close Encounters of the Fourth Kind: Alien Abduction, UFOs, and the Conference at M.I.T.* (New York: Knopf, 1995), 102.

10. Sympathy for the Alien

1. Philip K. Dick, *In Pursuit of Valis: Selections from the Exegesis*, ed. Lawrence Sutin (Novato, Calif.: Underwood-Miller, 1991), 7.

2. On the role of information transfer in abduction, psychologist John Mack writes, "The forces involved in the implanting, storage, and recovery of information remain among the central mysteries of the whole abduction phenomenon" (*Abduction: Human Encounters with Aliens* [New York: Ballantine, 1994], 52).

3. DJ Spooky, *Necropolis: The Dialogic Project*. Knitting Factory, 1996. Sampled from William Burroughs, *Nova Express*.

4. Parliament, *Mothership Connection*, Polygram Records, 1976.

5. Merlin Donald, *Origins of the Modern Mind: Three Stages in the Evolution of Culture and Cognition* (Cambridge, Mass.: Harvard University Press, 1991).

6. Report of Meetings of Scientific Advisory Panel on Unidentified Flying Objects Convened by Office of Scientific Intelligence, CIA. TAB A. January 14–18, 1953 (http://www.cufon.org/cufon/robert.htm).

7. Allen Hynek, *The UFO Experience: A Scientific Inquiry* (Chicago: H. Regnery Co., 1972), 22.

8. Ibid., 70.

9. Elias Canetti, *Crowds and Power*, trans. Carol Stewart (New York: Farrar, Straus, and Giroux, 1962); and Deleuze and Guattari, *A Thousand Plateaus*.

10. Hynek, 26 and Claude Shannon, *The Mathematical Theory of Communication* (Urbana: University of Illinois Press, 1949).

11. Edward Fredkin, *Finite Nature*, available at http://digitalphysics.org/Publications/Fredkin/Finite-Nature.

12. "Sammy Hagar: The Interview" at http://www.mtv.com/bands/archive/h/sammyfeature.jhtml. See also "Sammy on Politically Incorrect," http://www.redrocker.com/pi3/pi3pics.html. For another of Hagar experiences of sudden interconnectedness, see http://www.mtv.com/news/articles/1429852/19970530/story.jhtml. Finally, for a remarkable allegorical practice of body sampling, see http://www.livedaily.citysearch.com/news/982.html, where it is revealed that "Sammy Hagar will cut his red tresses on NBC's 'The Tonight Show with Jay Leno' on Nov. 12, [1999] and will donate them to a charity that makes hairpieces for children who lose their hair during medical treatments, according to Hagar's official website."

13. See Brian Rotman's treatment of the materiality of computation in *Ad Infinitum*. For a short introduction to the problem, see also Rolf Landauer, "Information is Physical," *Physics Today* 44 (May 1991): 23–29.

14. John Mack, *Abduction: Human Encounters with Aliens* (New York: Ballantine Books, 1995).

15. Whitley Strieber, *Communion* (New York: Beech Tree Books, 1987).

16. Jacques Derrida, *Margins of Philosophy*, trans. Alan Bass (Chicago: University of Chicago Press, 1984), 317.

17. John G. Fuller, *Incident at Exeter/The Interrupted Journey* (New York: MJF Books, 1966).

18. Dick, *A Scanner Darkly*.

19. Richard Doyle, "Dislocating Knowledge, Thinking Out of Joint: Rhizomatics, C. elegans, and the Importance of Being Multiple" *Configurations* 1 (1994): 47–58. Cited in Alan Sokal, "Transgressing the Boundaries: Towards a Transformative Hermeneutics of Quantum Gravity," available at http://www.physics.nyu.edu/faculty/sokal/transgress_v2/transgress_v2_singlefile.html.

20. Kip Thorne, *Black Holes and Time Warps: Einstein's Outrageous Legacy* (New York: W. W. Norton, 1994), 484.

21. Carl Sagan, *Contact* (New York: Pocket Books, 1997).

22. Dick, *Valis* (New York: Bantam Books, 1981), 23.

23. Ibid., 22.

24. Ibid.

25. Dick, *In Pursuit of Valis*, x–xi.

26. Dick, *The Man in the High Castle* (New York: Vintage Books, 1992).

27. Edward Fredkin, *Finite Nature*, available at http://digitalphysics.org/Publications/Fredkin/Finite-Nature.

28. Ibid.

29. Philip K. Dick, *The Shifting Realities of Philip K. Dick*, ed. Lawrence Sutin (New York: Vintage, 1995), 322.

30. Ibid.

31. Dick, *Shifting Realities*, 328.

32. Avital Ronell has written of this ontology in terms of beings on drugs dwelling in an ecology of expropriation. Her work is exemplary for and instructive to those of us attempting to hack ethical communities capable of responding to the futural qualities of informatic ecologies. She offers prescriptions—such as tropium—that are in fact practices of expropriation that remain in contact with a thoroughly differentiated other. See *Crack Wars: Literature, Addiction, Mania* (Lincoln: University of Nebraska Press, 1993).

33. Dick, *In Pursuit of Valis*, 2.

34. Ibid., 3.

35. Ibid., 6.

36. Deleuze and Guattari, *A Thousand Plateaus*, 238.

37. Allen Hynek, *The Hynek UFO Report* (New York: Barnes and Noble, 1977).

38. Mack, 389.

39. Nor can narrative, abducted, tell the truth of the future. A speech act referenced by abduction researcher Budd Hopkins, a remarkable command performance of the alien allegedly staged for Lech Walesa near the Brooklyn Bridge, stages sovereignty's—that political form that knows the future, it will have centers there—demise. This demise does not mark the end of politics, but the beginning of community for whom the command "Take me to your leader" is quite literally non-sense. Budd Hopkins, *Witnessed: The True Story of the Brooklyn Bridge UFO Abductions* (New York: Pocket Books, 1996).

Index

Richard Doyle is associate professor of rhetoric and science studies in the
Department of English at Pennsylvania State University. He is
the author of *On Beyond Living: Rhetorical Transformations of the Life Sciences*,
and he is currently working on the prequel to *Wetwares*,
tentatively titled *LSDNA: Psychedelic Science in America*.